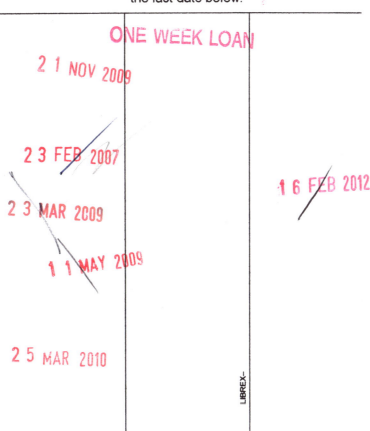

JCT 2005 – What's New?

James Davison

Acknowledgements

Contract clause headings and numbering structure from the:

- *Standard Building Contract With Quantities*
- *Design and Build Contract*
- *Intermediate Building Contract*
- *Minor Works Building Contract*

by the Joint Contracts Tribunal Limited, 2005, Sweet and Maxwell, © The Joint Contracts Tribunal Limited 2005, are reproduced here with permission.

Please note: References to the masculine include, where appropriate, the feminine.
Published by RICS Business Services Limited
a wholly owned subsidiary of
The Royal Institution of Chartered Surveyors
under the RICS Books imprint
Surveyor Court
Westwood Business Park
Coventry CV4 8JE
UK

ISBN 1 84219 231 0

Typeset in Great Britain by Columns Design Ltd, Reading
Printed in Great Britain by Cromwell Press Ltd, Trowbridge, Wiltshire

Contents

Preface

I have tried to keep the text practical and comparative rather than add to the number of tomes that already explain how JCT works. I felt that the person who would buy this book would already know JCT 1998 and when addressing the 2005 contracts for the first time would want to know what the differences were and what consequences the changes had had – if any.

I also thought that this imaginary reader would doubtless be in a hurry (and probably on the way to a meeting on site) as either the architect, contract administrator, QS, commercial manager or possibly even as a legal advisor.

For those reasons, the four most important JCT contracts were chosen. This was also the reason for providing an easily digestible overview of each contract before diving into the clause-by-clause analysis. I think the tabular format and the tables of destinations will help when readers need to get to the heart of the matter quickly.

In my opinion the new JCT 2005 contracts are much better than the 1998 forms they replace. Irrespective of this, I hope that this book ensures an even easier transition to the use of the new forms, so that the contracts don't get in the way of the actual business of constructing the house, hospital or shopping centre etc. that needs to be there at the end of the project.

Finally, some acknowledgements.

Firstly, I would like to acknowledge the contributions and support of my colleagues at Cyril Sweett:

- Tim Tapper, the director who leads Legal Support for Cyril Sweett: not only was he good enough to let me have the time away from the office to write but he also gave much of his free time to contribute to the editing of the book.
- Len Stewart gave me the first critical review of the book and contributed to the text to ensure that it remained practical and focussed on what actually happens on site.
- Andrew Hemsley for helping me order my thoughts and plan the task of tackling four brand new contracts.
- Steve Roberts and Sheila Peacock: certainly for their assistance in the editing and proof reading, but particularly for their work putting the tables of destinations together.

To the publishing professionals: Sabrina Kuhn and Ruth Wilkie at RICS Books and Jennifer Cowan at Etica Press, this is my first book and I have been grateful for your expertise and good humour throughout the project.

To my friends and family: apologies to those of you who put up with my absence in 2005 and many thanks to those of you with practical suggestions. In particular I would like to acknowledge my father, Chris, who practises as a QS in the North East and my brother, Paul, who practises as an architect in San Francisco.

Finally and most importantly I would like to thank Sophie Barrett-Brown, Head of UK Practice at Laura Devine Solicitors (and my wife). A superstar lawyer: were she not so successful and dedicated to her career I would have not have had the free time in which to write the book!

James Davison
February 2006

1 Introduction

Intention and structure of this book

Who is it for?

This book is intended for users familiar with the existing JCT standard forms of contract, who do not require an introduction to the 'JCT methodology' but need to know what has been changed in the new forms published in 2005. Such a reader will want to know what the changes mean in practice when using the new forms, where individual clauses have gone and how they have been changed. As such, the book has been written with contract administrators, architects, clients, quantity surveyors and contractors in mind, together with those who advise them.

The forms considered

The four forms considered to be the most important of the new 2005 JCT forms have been analysed in this book. These are:

- **Standard Building Contract With Quantities 2005 edition (SBC/Q 2005)** – this contract replaces JCT Standard Form of Building Contract 1998 edition 'Private With Quantities' *and* the 'Local Authorities With Quantities' 1998;
- **Design and Build Contract 2005 (DB 2005)** – this contract replaces JCT With Contractor's Design 1998;
- **Intermediate Building Contract 2005 (IC 2005)** – this contract replaces JCT Intermediate Form of Building Contract 1998;
- **Minor Works Building Contract 2005 (MW 2005)** – this contract replaces JCT Agreement for Minor Building Works 1998.

What is its purpose?

Each of the contracts considered contains new elements, and thus this book will also help when tackling the new processes, addressing new obligations and deciding which of the new options should be adopted (and, by extension, which ones are better left alone for a particular project). The book provides analysis and practical interpretation of each of the clauses of the four contracts that have been reviewed. It has been prepared as a point of reference for readers when they are using the new contracts and need to consider whether clauses in the new version have the same effect as they did previously under the JCT 1998 form.

This book aims to cater to that need by following each of the new contracts clause-by-clause and providing comparison, analysis and comment against each clause. A tabular format has been adopted to present this information for ease of reference so that readers can go directly to whichever clause or section they are concerned with.

A number of the clauses in the new contracts have the same form and substance as those of the previous forms of contract. Whilst there may be no change, this clearly does not mean that the clause can be ignored, but it does mean that users can rely upon their existing experience in addressing it.

Structure of the text

The book is divided into five chapters. This first chapter sets the background to the 2005 JCT contracts and considers some of the external factors that have influenced the drafting. It then provides a summary of the main changes found in the 2005 contracts, states general themes and notes missed opportunities.

The four main chapters each review one of the new forms. The first part of each of these chapters introduces the new form and considers the contract it is replacing. There is a summary of the most significant of the changes found in the new forms, and these are set in context.

The second part of the four chapters contains the main body of the text and provides the detailed analysis of the new forms. It is structured as a table in which each of the new forms is compared directly against its predecessor in an order that follows that of the new contract. In each instance, comparison of each clause is made with the relevant clause or clauses of the JCT 1998 form that has been superseded. Where there are no changes, this is indicated. Where there is a change, its order of importance and the potential consequences are stated and the new actions required as a result are identified.

The final part of these chapters contains a 'table of destinations'. Each table of destinations follows the structure of the old JCT 1998 and states the new clause reference in the 2005 form. The intention is that readers can use their 'JCT 1998' knowledge to help them find the relevant clause in the new contracts and find out where old clauses are not maintained in the new forms. It is therefore possible to travel in either direction – i.e. use a JCT 1998 clause to locate the relevant clause in the 2005 edition using the table of destinations in the third part, and vice versa using the table set out in the second part.

Why those four contracts?

The four forms chosen have been selected on the grounds that they represent the most significant share of the construction market in terms of the sums spent against them; incidentally, they are also the best-selling editions of the contract. Forms that have been published for the first time in 2005 have not been reviewed; thus the Framework Agreement and Minor Building Works Contract with Contractor's Design are beyond the scope of this work – not least as there is no direct predecessor against which the new contracts can be compared.

The background and influences on the new forms

Background

The Joint Contracts Tribunal (JCT) has ended its long association with the Royal Institute of British Architects' (RIBA's) publishing house and its standard form contracts are now being published by Sweet & Maxwell. More than 50 documents have been, or ultimately will be, published as part of this change and the new 2005 forms supersede the JCT's previous 1998 forms of contract. A new 'Digital Service' has been launched alongside the hard copy format.

JCT has taken the opportunity to update and amend its most important contracts, publish several brand new forms and retire or merge others.

The changes that have been made to the standard forms go beyond the new front cover, layout and wording: there are numerous important departures from the familiar policy positions established by JCT and these will catch unwary users out.

The new forms have a contemporary appearance and they are all set out in clear, well-ordered text that is divided into sections. The sections themselves are sensibly subdivided and the extensive use of titles aids comprehension. The new forms also have a high degree of consistency, and a similar layout and terminology applies across the separate forms of contract, which aids comparability.

In addition to the changes to the layout, there has been a general change to the text that seems to reflect a genuine effort to simplify the expression of the clauses, use plain English and limit the use of legal terms and archaic words.

To put it simply, the new forms look modern and compare favourably with their contemporary competitors. Going beyond the superficial, the contracts themselves are far more integrated and now have a greater degree of mutual compatibility and overall are better contracts than the documents they replace.

When one considers the results of the RICS Contracts In Use Survey of 2001, which found that JCT contracts are used in 90% of all construction projects, it is clear that the JCT's 2005 contracts will be of great significance to a sector of industry responsible for 10% of the United Kingdom's GDP.

Each of the forms reviewed has been improved as a result of the changes made for the 2005 publication of the forms. The changes that have been made to the substance of the contracts do not fundamentally recast the nature of the contracts, but do develop them. The changes that have been made have been made judiciously rather than capriciously. In general, the JCT drafting committee has, thankfully, resisted the temptation to jump on passing bandwagons and has maintained common sense, whilst improving the presentation, drafting and mechanisms of previous editions, and this should benefit all users.

Before considering the external influences that have helped to shape these new forms of contract, it is worth noting that there are very few reasons why it would be appropriate to favour procurement under a 1998 form of contract now that the 2005 forms have been published.

External influences on JCT 2005

Success of rival forms of contract
Since JCT 1998 was issued, other forms of contract have been issued that place certain procurement techniques at their heart. An example would be the Project Partnering Contract of 2000, a form which provides a thorough partnering-centred process which emphasises the need for early involvement of the contractor and the creation of a single project team. A new JCT Framework Agreement for 2005 provides the JCT's response to this procurement method.

The Engineering and Construction Contract (the 'ECC', also known as the New Engineering Contract or 'NEC') also struck out in a different direction to the other standard forms, including the ICE form of contract. The ECC requires unequivocal action to be taken at defined points in time to a far greater extent than any JCT contract and the ECC Third Edition (also published in 2005) continues this tradition.

This administrative bias is both the virtue and the vice of contracting under ECC. If followed correctly, it provides a structure and a process for good project management. However, if it is not applied correctly in practice then certain processes will be extremely difficult to resolve.

The ECC benefited from a sound layout that applied to each of its variant forms. The JCT forms have adopted a similar approach for the 2005 release. Although the wording of the ECC forms is at times difficult to apply, and although it does adopt an unusual style of drafting, it is written in modern English and has a limited reliance on overly legalistic language.

It is fair to say that JCT has not had to withstand competition that has been as sustained or effective as that represented by the various forms of ECC contract. This appears to have forced JCT to raise its game.

The format of JCT 2005 has been greatly improved and takes account of several of the developments made in the ECC forms. The 2005 forms have all had a general update in the language used and the JCT has been largely successful in its effort to reduce unnecessary verbiage. The JCT 2005 contracts have better document structure and layout in comparison to the 1998 form. Although JCT is fundamentally different to the ECC, there are substantive changes that echo ECC requirements; an example would be the requirement for increased clarity in timely decision making. If ECC has in any way influenced these changes, then JCT users can only be thankful for the improvements.

Rise of mediation

Mediation is a dispute resolution process that can produce solid results swiftly. The greatest attribute of mediation is that it leaves room for the Parties to resume a project without resorting to the entrenched positions that litigation, arbitration and adjudication necessitate. No resolution can be imposed in mediation; the process relies upon consensus being reached by the Parties.

Any dispute involving a construction project that is still on site, or where the Parties are likely to work together again, must give consideration to the mediation process as a way of resolving the problem. The process is gaining adherents, and is supported by the rules of the courts that provide great encouragement to mediate disputes.

The JCT drafting committee appears to have heeded this and the 2005 forms elevate the relative importance of mediation. A statement that mediation ought to be considered is the very first condition stated in the dispute resolution clause – the 1998 forms only made reference to mediation in the footnotes.

Changes to dispute resolution practice

At the time of writing, adjudication is in most instances the only formal dispute resolution process that will be applied to a construction dispute and the adjudicator is usually the only impartial third party to review the dispute. Cases moving on to final resolution based on full evidence before a judge or arbitrator are the exception. The impact of this is that a supposedly temporarily binding decision of the adjudicator is usually the final word on the matter.

The High Court Judge in charge of the Technology and Construction Court (TCC), His Honour Judge Jackson, observed in mid-2005 that the most common claims brought before the TCC concern the enforcement of adjudicators' decisions. Such cases concern procedural matters or questions approaching administrative law.

It will be interesting to see if there is any significant change in the volume of litigation in the TCC now that the JCT contracts have replaced arbitration with litigation as the default dispute resolution mechanism. However, the compound effect of the Woolf reforms and the *Housing Grants, Construction and Regeneration Act* 1996 is that vastly fewer construction cases go to court than before; of these, a small percentage involve substantive law or matters of interpretation.

Although the number of cases that went to court was always a small percentage of the total number of construction projects, the decisions that were produced gave guidance and certainty as to the interpretation of the clauses of the contracts and the application of new statutes.

The judges of the TCC are certainly more engaged in dialogue with the construction industry than before, in that they have contributed to the recent adjudication review, produced a second edition of the guide to the TCC and even had a number of articles published in the weekly *Building* magazine. Notwithstanding this, the shortfall in contemporary authority on the application of new statutes to new forms of contract does reduce certainty in the application of those forms. This increases the reliance on commonwealth authority or Scots law judgments on equivalent legislation. A further result is that the 2005 standard forms appear to have been less influenced by decisions of the courts in terms of the incorporation of specific amendment to accommodate specific decision than previous editions have been.

Acceptance of new procurement techniques

The JCT's drafting committee appears to have acknowledged the virtue of several contemporary procurement techniques. A good example is the Contractor's Design Submission Procedure in respect of DB 2005 and the Contractor's Designed Portion for the SBC/Q 2005.

It is appropriate for developments and trends to be adopted by standard forms. JCT has traditionally been prudent in offering such mechanisms as options or as completely separate contracts, rather than automatically writing new techniques into their standard forms.

This method has been repeated for the present release of contracts. For example, the publication of a new Framework Agreement (FA 2005) and a Minor Works Building Contract With Contractor's Design to provide for needs beyond those of the mainstream in separate new contracts. Equally, many of the innovative changes introduced for the first time in 2005 are options that must be specifically written in to the Articles in order to take effect.

In general, if users wish (or if they are obliged by policy) to deploy some of the cutting edge forms of procurement, then these objectives can often be accomplished by selecting the correct form of JCT contract and involving the contractor and professional team as early in the process as possible. Alternatively, bespoke amendments can be accommodated to achieve a satisfactory resolution.

Standard form users frequently want a known quantity upon which reasonable estimates of time and cost can be made. Construction is already sufficiently technically complex and prone to genuinely uncertain outcomes without having to work with wholly new forms of contract every few years.

It is far better for radical change or new techniques to be catered for in separate contracts or in stand-alone supplements than for them to become default positions in mainstream documents. If, in time, such techniques prove effective and become standard practice, then they

may well be catered for as options within the principal forms. Perhaps in the future we will see an optional 'partnering schedule' appended to a future edition of the JCT's Standard Building Contract.

Summary of critical changes

Changes of form or style

Layout and numbering
Each of the contracts reviewed has been restructured. The Articles of Agreement are found at the beginning of the contract and these contain the contract specific information in the Recitals, Articles and the Contract Particulars. There is no separate appendix or schedule of contract-specific information as there was in most of the 1998 forms.

The Conditions have been restructured and grouped into sections – in many instances, the running order of the clauses has been changed. The first consequence of this is that familiar clauses will need to be tracked down, and standard form documents that refer to clause references in the 1998 form or contract will need to be updated to reflect both the new numbering and the new terminology. The tables of destinations found in the third sections of the subsequent chapters of this book will assist in this task.

New terminology
A number of very familiar JCT phrases have been amended in the 2005 editions of the standard forms. Examples include the change from the familiar term 'Defects Liability Period' to 'Defects Rectification Period'; similarly, the old term 'determination' has been changed to the more contemporary term 'termination'. Certain certificates have been renamed – for example, the 'Certificate of Completion of Making Good Defects' is now simply the 'Certificate of Making Good'.

In addition, an attempt has been made to correct terminology in the 1998 edition that was perhaps over-specific. For example, both the 1998 forms and the 2005 forms permit a change to the completion date to take account of matters beyond the control of the Contractor and also to take account of work being taken away from the Contractor, with the result that the period of time for completion of the Works could, within certain restraints, either increase or be reduced. Both situations are now covered under the heading 'Adjustment of Completion Date' rather than 'Extension of Time'.

These changes reflect a general trend to update the words and the terminology of the 2005 contracts, but the changes do mean that certain assumptions about the meaning and application of the terms will need to be reassessed. Certainly standard documentation such as certificates and tender documentation will have to be updated to take account of the new terminology.

Incorporation of supplements
Several of the more popular supplements that JCT published as separate documents are now included within certain of the standard forms. For example, SBC/Q 2005 gives the user the option for Sectional Completion, Collateral Warranties and a Contractor's Designed Portion (CDP).

It is still necessary to activate these clauses specifically and consider how they interrelate, but they are now effectively coordinated with each other and it should not be necessary to draft enabling clauses when both CDP and Sectional Completion options are used.

The avoidance of the practical administrative inconvenience of sourcing supplements and the associated delay will also be of value.

A true Standard Building Contract
Previous editions of the JCT contract have involved the publication of separate forms of contract for employers in the private sector and the public sector. The 'local authorities' versions of the JCT contract took account of the limitations placed on the actions of local authorities. The 2005 Standard Building Contract seeks to cater to the requirements and capacity of both public- and private-sector employers and no separate 'private' or 'local authorities' version will be published.

The consequence of the merging of these two forms is relatively benign, in that neither private- nor public-sector employer appears to be disadvantaged by the change. Generally the distinction in the status of the differing kinds of employer is accommodated by the activation or suspension of certain clauses of the SBC/Q 2005 depending on whether or not the Employer is a local authority. However, there are some changes of terminology – for example, the term 'Architect/Contract Administrator' has been adopted throughout SBC/Q 2005. This would appear to adopt the term used in the local authorities version of the 1998 form to accommodate the fact that the Contract Administrator may well be an officer of the authority.

Changes of substance

Insurance for Contractor's Design Liability
As mentioned above, the supplement providing for a portion of the Works to be designed by the Contractor has been integrated into SBC 2005. Furthermore, both SBC 2005 and DB 2005 include provisions requiring the Contractor to take out Professional Indemnity insurance in respect of design liability. Note that the Parties must specifically opt-in for the clause to take effect, but this is a significant development as, surprisingly, neither previous forms, nor the Contractor's Design Portion Supplement, made such provision.

Submission of Contractor's Design

DB 2005 adopts and adapts an effective Contractor's Design Submission Procedure from the 2003 Major Projects Form. The timescale for action by the Employer is tight and in the absence of any action by the Employer, the Contractor is to proceed on the basis of his designs. In practice, the Employer's vetting process is likely to be carried out by the Employer's Agent or a design consultant that the Employer has retained specifically to carry out this function.

The procedure should leave no doubt as to which design documents can be used for construction, but it does not result in any transfer of design liability. Thus the Contractor's obligations are not diluted by compliance with the procedure.

Dispute resolution

The change to the default dispute resolution model is a major change. Although the forms provide an option for all disputes to be arbitrated, arbitration is *not* the default position. Instead litigation is the default dispute resolution procedure. This is a substantial change and the Contract Particulars would have to be amended to restore arbitration to the default option if this was the favoured option.

The status of adjudication has not been changed, but the 2005 forms use a different set of adjudication rules to the 1998 forms. The 1998 forms made use of the JCT's own adjudication rules which were set out in the conditions. The 2005 forms make use of the Scheme set out in the *Scheme for Construction Contracts (England and Wales) Regulations* 1998 ('the Scheme'). The Scheme has been amended to allow the selection of a specific Adjudicator Nominating Body in the Contract Particulars and to require the involvement of a specialist if the substance of the dispute involves testing.

The 2005 forms also recognise the increasingly apparent merits of mediation, and the new forms contain a clause that presents the mediation of disputes as a valid method of settling disputes. Whist this clause is not binding, it may well make sound commercial sense to consider a mediated settlement instead of resorting to the (necessarily) confrontational options otherwise provided.

Payment notices and withholding payment

The statement of the requirement for a notice concerning what is going to be paid and the requirement for a withholding notice have been sharpened. Both should be issued, but one notice or either notice in sufficiently precise terms and with all the requisite details will suffice, if served at the correct time.

Failure to issue payment or withholding notices under DB 2005

A change of particular significance which is found in DB 2005 specifically reverses the position found in WCD 1998. If the Employer fails to issue a payment notice or withholding notice under DB 2005, the Contractor will be entitled to be paid the sum due, unlike WCD 1998 which provided for the payment of the sum the Contractor requested.

Retention

The standard level of retention has been reduced from 5% to 3% for SBC/Q 2005 and DB 2005. Retention remains 5% for the Minor and Intermediate forms.

Third party rights

The mechanism for providing rights to third parties has been revamped for the 2005 release of contracts.

There are two methods of providing rights to third parties such as Funders, Purchasers and Tenants.

The first is the traditional method of providing collateral warranties in favour of the third party and the JCT's standard form collateral warranties are now included in DB 2005 and SBC/Q 2005 (instead of being separate supplements). The conditions require that the collateral warranties are provided within a defined number of days of the request being issued.

The second method is an innovation and, if accepted in practice, will obviate the need for collateral warranties. JCT 2005 provides for the first time that Parties may make use of the *Contracts (Rights of Third Parties) Act* 1999. The Parties do this by describing the third parties and stating the rights that they are conferred. This option does not require the (frequently painful) task of getting the warranties executed.

The default position is that there are to be no rights for third parties. If third party rights are required then Part 2 of the Contract Particulars must be completed to enable whichever of the two mechanisms is favoured.

Notice required for termination for insolvency
The contract may still be terminated for the insolvency of a counterparty, but the new forms require a notice to be issued to the defaulting Party to achieve the termination. Although there were exceptions, in almost all instances the termination occurred automatically upon insolvency under the 1998 forms.

The consequence of this is that a definite action, in the form of the issue of a notice, must be taken by a Party wishing to terminate a contract following the insolvency of its counterparty.

Users should also note that the definitions of insolvency have been adjusted. On balance, the threshold has been lowered and acts that might previously have been permitted may now amount to insolvency under the new definition.

Security for advance payments
There is no requirement for an advance payment to be made to the Contractor. In the event that the mechanism is adopted then the 2005 contracts now automatically require the Contractor to provide an Advance Payment Bond. This is a shift in the default position in the event that the option is selected – the terms of the Advance Payment Bond itself are fundamentally unchanged.

Adjustment of Completion Date
There are a number of detailed but highly significant changes in relation to the process of adjusting the Completion Date.

The Architect/Contract Administrator (ACA) still has a 12-week period in which to respond, but the approach of the date for completion no longer undercuts that period as the duty is only to endeavour to complete the assessment before the date for completion – previously this was a requirement.

The role of the ACA (or the Employer if it is DB 2005) has become more exacting and the response expected of the ACA to a notice of a Relevant Event is now more precise.

The ACA must now state what award of time is given against each of the Relevant Events notified. This is an exercise of some technical complexity that requires the identification of matters that are not considered to be Relevant Events, as well as Relevant Events where no time is due. What is likely to be required is an itemised statement that does not bundle the Relevant Events together and state a general period against a number of Relevant Events.

This is a higher standard of detailed response than was required by the 1998 forms. Although many ACAs presented their responses in such a manner under JCT 2005, it might be argued that the ACA will need to spend more time in the preparation of a specific response to each Relevant Event notified by the Contractor.

Relevant Events

The grounds for adjusting the completion date have been reviewed – several provisions relating to labour and equipment have been deleted (as almost inevitably happened in practice) and a single catch-all term is used to sweep up several Relevant Events that were previously stated in their own clause.

Parity in suspension

The 1998 forms stated different time periods in respect of a suspension giving rise to a right to terminate. Put simply, the Contractor had to wait longer before he was permitted to treat suspension as a termination event. The 2005 forms remove the distinction and a new default period of two months is stated.

No nomination provisions

DB 2005 and SBC 2005 no longer make provision for nominated sub-contractors. It should be noted that the nominated sub-contract mechanism has been retained for IC 2005.

Two general trends

From an analysis of the changes made to the JCT 2005 edition two general trends can be discerned, as discussed below.

A higher standard of contract administration

As a general point, a greater level of clarity and decisiveness is required of an Architect/Contract Administrator (ACA) under each of the 2005 forms.

This is exemplified by the requirement for the ACA to provide a specific response in respect of every notice and for every Relevant Event. There is also a general suggestion that the ACA must get on with making the assessment rather than entering a cycle of requests for information.

In this respect, the changes to the process when adjustments to the completion date are considered have the effect of reducing the opportunity for delay and equivocation of both the ACA and the Contractor.

The need to state clear and unequivocal decisions in writing and in detail needs to be considered against the rights of Parties to seek adjudication of disputes. Clarity in each contract administration decision defines a matter that might be validly disputed. This needs to be set against the fact that a clear expression will also place certain limits upon the scope of the dispute and will tend to reduce general and widely defined disputes. On balance, this is a positive development.

Greater clarity at the outset

This clarity may be seen in two specific respects. Firstly, the contracts are more complete in their standard form – there is less of a requirement to purchase separate supplements or attachments.

Secondly, the statement of a thorough set of Articles of Agreement, including a detailed set of Contract Particulars which resolve to default positions rather than blank spaces, reduce the uncertainty as to the base position of the contract.

The consequence of this clarity is that decisions to accept the default positions or take up the options should be made sooner in the procurement process. It is expected that this will enable a reduction in the time it takes to produce contract documents for tender and subsequent execution. This will also help when the Parties do not manage to complete the contract or forget to complete an appendix etc.

Missed opportunities

Although the improvements that have been made to the 2005 forms of contract certainly represent a genuine improvement in the quality and completeness of the JCT's standard forms, there are certain changes that might have been made, but have not, and amount to missed opportunities.

Standard Novation Agreement for Design and Build Works

There is no JCT Standard Form of Novation Agreement. There is also no form of appointment for a designer (be it an engineer or architect) that is drafted with the specific intent of the employment of that professional being transferred from the Employer to the Contractor, at some time during the term of the appointment.

At present many 'novations' actually involve the renegotiation of contract terms. A standard JCT contract that addressed the processes to be applied to actually regulate the transfer on conditions established at the outset would reduce uncertainty at the point of transfer for all concerned.

The main merit of such a standard document would be the certainty that Employer, Contractor and Designer would each gain at the outset of the project. A clear statement of the obligations after the transfer would protect both the Contractor and the Designer.

Given the fact that selection, appointment and subsequent transfer of a design team are utterly standard procedures, one wonders why a standard form has not been developed. It is accepted that such a form of professional appointment or novation agreement would reduce the commercial latitude of the Parties at the time of transfer. Perhaps

Parties might consider that the present arrangement enables them to adapt to changes that occur between the initial design phase and the construction phase. It is suggested that this would be to ignore the fact that a contract addressing the transfer would aid the appreciation of the legal and commercial proposition that both designer and Contractor must consider when approaching the project in the first place. Such a JCT standard document could specifically address design liability and by its terms state how such liability should – or for that matter how it should not – be transferred.

Pre-construction agreement

Construction projects are often bedevilled by the disconnection that can develop between the project on paper and the project on site. Projects are often well underway on site, often for sound construction or commercial reasons, long before the terms and conditions are executed. The expedient work-around, so frequently used, is the letter of intent. At the risk of reciting a truism, a great many projects get underway and approach completion of the construction phase solely on the strength of a letter of intent without the main contract ever being executed.

Many projects experience the transition between the drive to execution of the main terms and conditions of contract and rush to get something in place to get the works underway on site. In most instances the letter of intent is in a bespoke or in-house style that will create certain contractual rights and obligations and may fall within the scope of a written contract for the purpose of the *Construction Act*. Such letters of intent will usually seek to apply the terms of one of the JCT contracts until such time as the full agreement is executed.

In fairness, the JCT's 2005 forms address this problem more comprehensively than most other standard forms by providing default positions on a number of the substantive matters that might otherwise cause a contract to founder for uncertainty – or that might at least cause uncertainty at the most inopportune moments.

However, this does not get around the fact that projects will continue to get on site, even under the 2005 forms, long before the terms are finally agreed: this presents a risk for all involved.

Were the JCT to produce a standard commencement agreement, pre-construction agreement or even a standard form set of heads of agreement this would not only grease the slipway for the main terms and conditions but would also provide an established legal certainty as to the respective obligations and the limits on those obligations for all Parties. It would be expected for such a document to be thoroughly tied to the obligations of the main set of JCT terms and conditions.

If such a document were produced, then it is accepted that there is a risk that the execution of an effective commencement agreement might be achieved at the expense of the conclusion of the agreement of the main terms and conditions. It is suggested that such a risk be set against the hazards of proceeding on a letter of intent that may be insufficient. For most projects this will be a small risk, worth taking in comparison to the potential difficulties that will arise in the absence of any useful agreement.

A standard form JCT commencement agreement or pre-construction contract would be a great benefit to all those concerned in putting

projects together and, by stating the material obligations would significantly reduce the number of disputes that would arise as a result of a failure to appreciate the obligations.

No objective definition of 'practical completion'

A definition of 'practical completion' was provided in the Major Projects Form of 2003, but no objective definition of practical completion is found in any of the 2005 editions of the contracts reviewed in this book.

It is understood that the reason for this is that the scope of projects that are procured under each form of contract is such that any standard definition would not accommodate the specific requirements of each project. In addition, it might be argued that an objective standard might well fail to capture the essentially subjective question of whether a building or section of a building is ready for beneficial occupation.

Although such concerns should be noted, the JCT 2005 forms would have benefited from a definition of practical completion – even a number of optional definitions with a default would have assisted, as this is an area that frequently causes problems.

2 JCT Standard Building Contract With Quantities 2005

Introduction

The JCT Standard Building Contract With Quantities 2005 edition (SBC/Q 2005) is the successor to the JCT Standard Form of Building Contract 1998 edition 'Private With Quantities' *and* the 'Local Authorities With Quantities' – no separate Local Authorities version of the form will be published. It can be considered to be the flagship document of the JCT 2005 release of standard form contracts, and if it is as successful as its predecessor, it is likely to be the most significant form of construction contract in the United Kingdom until such time as it is itself superseded by the next JCT contract.

Overcoming the problems with JCT 1998

It was said that despite its length and complexity JCT 1998 still failed to resolve many of the problems that arose on-site and in practice. It is unsurprising that a contract with as long a history and as great a weight of precedent as JCT 1998 was subject to some criticism. The contract suffered from an accumulation of clauses to resolve technical problems and market conditions that have passed, or to accommodate the rulings of the courts.

Although amendments are issued periodically by the JCT, these tend to incorporate new clauses rather than remove old ones that have fallen out of use. The result is that the contracts tend only to get longer and gain greater convolution in order retrospectively to graft on additional or optional processes.

The result is a contract that can be difficult to read and even anachronistic: this can make it difficult to apply, and unwieldy in practice. The lack of a definition of 'practical completion' is instance of JCT's history bearing down upon its present form. This is one of JCT's traditional oddities that has slipped through the net and remains unchanged and undefined in SBC/Q 2005, despite the fact that the JCT Major Projects Form of 2003 included such a definition.

There seems to have been a determined effort on the part of the SBC/Q 2005 drafting committee to rout out processes that have remained (or at least remained unchanged) through inertia. The best example of their efforts is the welcome removal of all the provisions for the nomination of sub-contractors in SBC/Q 2005.

Perhaps most significantly for most users on a day-to-day basis, is the fact that the drafting committee appears to have sought to redraw and reorganise the contract terms in plain English that is comprehensible to a modern reader and less reliant upon old fashioned terminology than its predecessors.

Many of the changes relate to layout and interpretation rather than to balance of risk, actions and obligations, however this should not be misinterpreted as a simple change in style alone. A clearly drafted contract allows a Party to know what is required of it and what actions it may take in any given circumstance. Well-drafted obligations are less susceptible to alternative interpretation that breeds dispute and the potential for litigation.

Purpose of the exercise

Having introduced the headline concepts in this opening part of the chapter, the main body of the chapter seeks to analyse the whole of the new contract clause by clause with the aim of identifying and

JCT 2005 Standard Building Contract
With Quantities (SBC/Q 2005)

explaining the differences between JCT 1998 and SBC/Q. This is a practitioners text for users familiar with the existing JCT 1998 forms of contract and the exercise purposefully avoids analysing clauses that have not been changed in the 2005 version of the contract.

The second part of this chapter has been set out in a tabular format to aid use as a point of reference when Contract Documents are being prepared and when the contract is used in practice on site. Reading from left to right the columns are:

- Column One: states the number and name of the SBC/Q 2005 clause;
- Column Two: states the number and name (where there is one) of the JCT 1998 clause;
- Column Three: a brief statement summarising any change and its nature. Standard bands indicate the degree of change at a glance;
- Column Four: examines the specific significance of any changes including a consideration of the consequences of the differences for the operation of the contract. As such this column is the heart of the book;
- Column Five: states new actions that the Parties must perform and, where relevant, provides guidance on what this means in practice.

Base documents
In this chapter, SBC/Q 2005 has been compared to JCT 1998 Private With Quantities incorporating Amendments 1: 1999, 2: 2000, 3: 2001, 4: 2001, 5: 2003 (this is referred to as 'JCT 1998' throughout the text).

In addition, comparison is made between SBC/Q 2005 and the following documents:

- Sectional Completion Supplement with Quantities 1998 (revised October 2003);
- Contractor's Designed Portion Supplement with Quantities 1998 (revised November 2003);

- Collateral Warranty from Contractor to Purchasers and Tenants (MCWa/P&T) 2001 edition;
- Collateral Warranty from Contractor to Funder (MCWa/F) 2001 edition.

Summary of the changes in SBC/Q 2005

SBC/Q 2005 is in part a consolidation of the amendments that have been published since 1998, but it is also the positive result of an effort to improve the usability and comprehensibility of the JCT contract, to make it more useful to its users. In general this has been achieved, and it appears that a significant proportion of the clauses have been redrafted to improve clarity without changing the intent or effect.

SBC/Q 2005 has a new layout and is written in plainer English than JCT 1998, but familiar processes have been retained. Although there are changes that mean that users will have to act differently when acting under SBC/Q 2005, it is based upon a traditional model of construction procurement and will be familiar to those already experienced with JCT contracts.

Significant changes

Changes of form

Use of titles and subheadings
SBC/Q 2005 is more thoroughly subdivided than previous JCT contracts and is arrayed into sections, each of which is described with short titles. The clauses themselves are frequently given titles but, like the section headings, the titles are for convenience only not for interpretation (clause 1.4.1).

Users should not expect clauses to have retained the same order. Even though the essential features of many of the clauses is unchanged, the 'running order' of the clauses has been substantially reviewed and the document is perhaps more readable as a result.

Notwithstanding the reordering of the clauses, the use of sections and clause headings throughout greatly assists when trying to locate clauses in the text and is certainly an improvement on JCT 1998, where one would need to refer to the small text off to one side of the main clauses.

Changes to terminology

A number of familiar terms have been revised. For example, the contract is now administered by an 'Architect/Contract Administrator' rather than by an Architect – this change reflects the terms used in the Local Authorities version of JCT 1998. There is now an 'Adjustment to the Completion Date' under SBC/Q 2005 rather than an 'Extension of Time'.

A number of familiar communications under the contract are now given names such as the 'Non-Completion Certificate' (clause 2.31). Other notable changes update the language; for example, the contract is now 'terminated' rather than 'determined'. Some of the changes simply reorder already familiar names; for example, 'Practical Completion Certificate' replaces the JCT 1998 term 'Certificate of Practical Completion'. Similarly the new term 'Certificate of Making Good' replaces the JCT 1998 term 'Certificate of Completion of Making Good Defects'.

The statement of a default position

All standard form contracts must address the conflict between the twin needs of establishing a single, certain and authoritative form of contract with a contract that can be tailored by users to suit their circumstances. This conflict lies against a background of contract law that permits Parties infinitely to tailor the terms of an Agreement (or amend a standard form) to the perceived needs of a particular transaction.

JCT 1998 represented a comprehensive whole, but presented a series of options to the user that required a selection to be made or perhaps required the user to buy further supplements. Previous JCT contracts placed the onus of selecting an option or incorporating a supplement upon the Parties and their advisors.

In certain key instances SBC/Q 2005 has moved away from this model and the contract provides a default position from which the Parties must specifically deviate. There are now significant points on which a default position is stated, and the prospect of the user being left with an incorrectly formulated contract is reduced as a result. Users should note that this is not the case for all of the choices available in the contract and a number of options must still be selected.

Providing a default position will not give all Parties what they require for all projects, but it does give a degree of certainty. The stating of the default is usually done in a way that does not prevent a project-specific requirement being stated in preference, or an amendment being inserted where necessary. The statement of a default position is certainly likely to assist in those cases (which are not particularly rare in construction!) where the contract ends up unexecuted but the Parties have adopted its terms.

In conclusion, this is a pragmatic change that will be of great use in practice. The default position is a base position that one anticipates the JCT believes to be fair following its consideration of law, practice and commercial reality, but it comes with the acknowledgement that particular projects and sophisticated users may have specific requirements.

Changes of substance

Payment process and retention
The basic payment system for the contract is unchanged, but is better expressed in terms of the way that the contract deals with the requirements for withholding notices and payment notices. The contract follows the *Housing Grants, Construction and Regeneration Act 1996* ('the Construction Act') and provides for two notices to be issued to the Contractor after the Interim Certificate has been issued by the Architect/Contract Administrator – one payment notice (clause 4.13.3) and one withholding notice (clause 4.13.4). If correctly put together (in terms of the statements of the sums to be withheld, with reasons for the withholding), then the payment notice may also serve as the withholding certificate, with the effect that no subsequent notice is necessary. Having served a payment notice the Employer may then serve a withholding notice, but he may serve this notice even if there has been no payment notice under clause 4.13.4.

The Parties should note that the standard sum for retention has been decreased from 5% to 3%.

Contract Particulars
This form seeks to place all of the project-specific information, (which users will frequently complete by hand), in one place at the front of the document in a new section called the 'Contract Particulars'. JCT 1998 placed much of this information at the end of the contract, but it was also spread throughout the Recitals, Articles and the main text of the contract. Grouping the project-specific information should assist in document preparation and in practice and is an improvement on JCT 1998. In addition, some of the information required will assist in the operation of the contract mechanisms – for example, the provision of an address for the service of documents.

Dispute resolution
There have been revisions to the dispute resolution process at each level. Most significantly, SBC/Q 2005 has changed the default position from arbitration to litigation. Previous JCT forms have sought to provide for the arbitration of all disputes under the contract.

The effect was such that cases before the court would be stayed pending the outcome of arbitral proceedings. SBC/Q 2005 does not provide for arbitration of all disputes arising under the contract unless the Parties specifically select it in the Contract Particulars. This means that the Parties are more likely to use the courts to resolve their disputes.

The adjudication process has been changed substantially and will now be carried out under the rules of the Scheme set out in the *Scheme for Construction Contracts (England and Wales) Regulations* 1998 ('the Scheme') rather than the JCT's own rules. Perhaps this departure from the JCT's own rules is in part a reflection of the authority that has built up to aid comprehension and application of the Scheme.

The contract also has a new clause that suggests that the Parties consider the mediation of disputes (a process that has found much favour in the courts).

Third party rights
The construction contract between the Employer and the Contractor is only the most obvious of the obligations a construction project puts in place. Users will be aware that the English law of contract only enforces contract obligation on the Parties to the contract and that a practice developed to accommodate the need for other Parties whereby sub-contractors, designers and Contract Administrators entered into additional contracts with Funders and Purchasers (usually described as 'collateral warranties'). The provisions of the recent *Contracts (Rights*

of Third Parties) Act 1999 sought to redraw some of the rules of privity of contract, but were usually excluded to the maximum extent possible.

JCT 1998 did not include terms for the provision of collateral warranties, nor did it provide standard forms of warranty – these were published as stand-alone supplements and had to be purchased and incorporated on a job-by-job basis.

Section 7 of SBC/Q 2005 requires the provision of a standard set of rights for the kind of third party that typically requires them due to a specific interest in construction activity rising out of the provision of finance or the purchase of assets. The rights themselves are set out in Schedule 5 and may either take the form of collateral warranties that are executed in documentary form, or alternatively by adoption of the *Contracts (Rights of Third Parties) Act* 1999. The provision of rights to third parties is not mandatory and must be specifically selected in Part 2 of the Contract Particulars, otherwise there will be no obligation for them to be provided.

The third party rights set out in SBC/Q 2005 provide a reasonable balance of the requirements of third parties with the obligations that can be accepted by Contractors, professionals and sub-contractors.

The contract provides a standard set of rights for a Funder; these apply throughout the construction period. It also provides step-in rights for that Funder. The rights of the Purchaser and Tenant (P&T) are more limited than those of the Funder, and only apply from practical completion onward.

In addition to the rights that the P&T or Funder may have against the Contractor, provision is made for collateral warranties to be provided by sub-contractors and professionals to third parties. The terms are substantially similar to existing JCT documents that had been previously published as supplements.

The incorporation of the JCT standard warranties and the obligations to ensure that they are provided is certainly welcomed, although this is tempered by the reality that Funders are uniquely placed to ensure that their particular requirements are met. It also remains to be seen whether Funders, Purchasers or Tenants will be satisfied to rely upon the *Contracts (Rights of Third Parties) Act* 1999.

Termination
The termination clause has been updated and the insolvency definitions changed – the most significant change is that the contract will not immediately terminate in the event of insolvency.

SBC/Q 2005 requires that a notice of termination must be served to bring a contract to an end due to the insolvency of a counterparty, whereas JCT 1998 was automatically determined upon the occurrence of most insolvency events.

Electronic communications
The Parties may agree that certain communication required by the terms of the contract can be sent by electronic means. If this is required, they must complete the section in the Contract Particulars. This amendment replaces the JCT 1998 scheme for an Electronic Data Interchange and has a number of implications. Of these, the most significant is that, if correctly drafted, the contract might provide that all communication and notices required under the contract could be sent electronically.

Many projects are to all intents and purposes managed in this way, with instructions, correspondence and even payment certificates being issued by email. It is often the case that terms and conditions require hard copy notices to be issued, and thus the electronic or faxed document arrives some days before the written notice finally arrives.

The risk of such change is that it downgrades the apparent significance of documents that have very serious consequences if they are not dealt with in an appropriate manner. The other risk is the perennial problem where a difficulty that has been elliptically alluded to in an email between staff is later said to represent good notice. It would seem that professionalism would safeguard against such consequences; provided that the staff drafting the correspondence do it correctly and state its purpose clearly, and that the staff receiving the correspondence are alive to the potential that each email might have, then quick and clear communication can ensure that the contract is operated more effectively.

Incorporation of supplements
The published edition of SBC/Q 2005 includes provisions that were previously addenda or separate supplements to the main contract. These are the Sectional Completion Supplement, the Contractor's Designed Portion Supplement and also the Collateral Warranties and Fluctuations; these were previously supplements but are now included in the standard document.

Parties are by no means obliged to adopt the sectional completion regime, nor is there any requirement for the Contractor to carry out design activity merely because clauses enabling such procedures are offered in SBC/Q 2005.

If the Parties do wish to make use of such techniques, their incorporation into the text is managed in a way that is effective. The accommodation of options and conditional events has the tendency to make all contracts more difficult to read and therefore more difficult to understand and apply. This was especially the case when an

appendix or supplement was being applied to previous versions of JCT contracts. In general, the incorporation of these options into SBC/Q 2005 and the text required to make it effective does not obscure the meaning of the clauses for those users who do not want to make use of the options.

An added specific benefit is that the clauses are coordinated with each other – it is no longer necessary to amend the contract to accommodate the use of sectional completion and a Contractor's Designed Portion on the same project.

Contractor's Designed Portion
Like JCT 1998, SBC/Q 2005 is a contract for which the Employer is expected to employ its own design team. However, unlike JCT 1998, SBC/Q 2005 is supplied with the facility for an element of design work to be carried out by the Contractor – the 'Contractor's Designed Portion' (CDP).

A key distinction for SBC/Q 2005 is that the Contractor is now obliged to obtain and maintain design liability insurance (clause 6.11), as insurance of this kind was not a requirement of the 1998 CDP Supplement.

The CDP is referenced throughout the contract where relevant, and the process introduced is an updated version of that set out in the 1998 CDP Supplement. The decision to make use of the CDP must be made early in the procurement process, and certainly before the contract is executed. It should not be taken up part-way through the contract process.

Sectional completion
A sectional completion regime can be of use to projects that have discrete elements that the Employer needs to be delivered at different

times. A classic example would be a shopping centre with Sections to be handed over to Tenants or opened in separate phases. The 1998 Sectional Completion Supplement has been largely adopted wholesale to provide this option.

Mandatory bonds

The contract provides additional security for employers providing an advance payment to the Contractor and adopts the default position that the Contractor is obliged to provide an Advance Payment Bond in respect of the sums advanced. The form of the bond is appended to Schedule 6 of the Contract. The provision that advance payment ought to be so secured may lead to an increase in the use of advance payments.

Changing the time for completion

The adjustment of the time for completion retains the integrity of the construction period and ensures that the Contractor will still be liable to pay liquidated damages if he fails to achieve practical completion on time.

The Relevant Events that may give rise to a need to adjust the Completion Date have been slightly updated for SBC/Q 2005. For example, the clauses that accommodated the economic difficulties of the 1960s and 1970s (in terms of labour relations and supply of materials) have been streamlined and are no longer stated as Relevant Events. The result is that the Contractor is more thoroughly responsible for his own supply chain (in terms of both labour and materials). This also reflects the present understanding of the services provided by Contractors and their bargaining power, as well as any wider economic significance that could be interpreted.

Equally importantly, the Architect/Contract Administrator must now state the reasons that he takes into account when adjusting the Completion Date. In this respect the new contract almost follows the withholding notice format where a reason and the consequence must be stated together – in the present instance, this is a period of time awarded (in the instance of a withholding notice it is a sum of money). Users, particularly contract administrators, should be careful to update their processes and ensure that this new provision is satisfied when adjusting the Completion Date under JCT SBC/Q 2005.

19

JCT 2005 Standard Building Contract with Quantities (SBC/Q 2005)

SBC/Q 2005 clause and reference title	JCT 1998 clause and reference title	Summary of change	Consequence of change and comments	New action required
Articles of Agreement			*Generally:* Although the Articles have been substantially derived from JCT 1998 they have been sharpened and the words used are more contemporary.	
Recitals				
First Recital (*the Works*)	First Recital	Slight amendment: the clause repeats the text of JCT 1998 with slight changes that improve the drafting or syntax, but the effect is unchanged.	Users should take more care under SBC/Q 2005 to complete the description of the Works than under JCT 1998. This is because the description now becomes the definition of the Works. A careless description could therefore conceivably commit the Employer to expensive additions thought to be included in the Works, or the Contractor to expensive additional work.	
Second Recital (*The Contract Bills*)	Second Recital	Slight amendment: the clause repeats the text of JCT 1998 with slight changes that improve the drafting or syntax, but the effect is unchanged.	Activity Schedule now to be completed by the Contractor and included as an annex to the contract.	
Third Recital (*The Contract Drawings*)	Third Recital	Slight amendment: the clause repeats the text of JCT 1998 with slight changes that improve the drafting or syntax, but the effect is unchanged.		
Fourth Recital (*The Construction Industry Scheme*)	Fourth Recital	Slight amendment: the clause repeats the text of JCT 1998 with slight changes that improve the drafting or syntax, but the effect is unchanged.		

Fifth Recital *(The Information Release Schedule)*	Sixth Recital	No change: the clause repeats the text of JCT 1998.		
Sixth Recital *(Sections)*	Sectional Completion Supplement	Slight amendment: the clause repeats the text of the 1998 Sectional Completion Supplement with slight changes that improve the drafting or syntax, but the effect is unchanged.	The division of the Works into Sections is not mandatory and remains an option that must be specifically selected. The difference compared to JCT 1998 is that such changes as are required to accommodate sectional completion are now found throughout the contract rather than contained in a separate supplement.	If sectional completion is required, then the contract will have to be prepared on this basis. Note also that if the Works are to be undertaken in Sections then most Certificates and notices will need to take account of this.
Seventh Recital *(The Contractor's Designed Portion)*	Contractor's Designed Portion Supplement	Slight amendment: the clause repeats the text of the 1998 CDP Supplement with slight changes that improve the drafting or syntax, but the effect is unchanged.	Recites the optional inclusion of a CDP. The CDP regime for SBC/Q 2005 is substantially the same as that found in the 1998 JCT CDP Supplement. The Employer's team must take particular care in the selection of the items that the Contractor will be asked to design and in assembling the Employer's Requirements. CDP is not mandatory – there is no requirement for users of SBC/Q 2005 to require Contractors to carry out design work.	
Eighth Recital *(the Employer's Requirements)*	Contractor's Designed Portion Supplement Third Recital and definition of 'Employer's Requirements'	Slight amendment: the clause repeats the text of the 1998 CDP Supplement with slight changes that improve the drafting or syntax, but the effect is unchanged.		
Ninth Recital *(Contractor's Proposals and CDP Analysis)*	Contractor's Designed Portion Supplement – Fourth Recital	Slight amendment: the clause repeats the text of the 1998 CDP Supplement with slight changes that improve the drafting or syntax, but the effect is unchanged.		

21

JCT 2005 Standard Building Contract With Quantities (SBC/Q 2005)

SBC/Q 2005 clause and reference title	JCT 1998 clause and reference title	Summary of change	Consequence of change and comments	New action required
Tenth Recital *(Employer's satisfaction with the Contractor's Proposals)*	Contractor's Designed Portion Supplement	Slight amendment: the clause repeats the text of the 1998 CDP Supplement with slight changes that improve the drafting or syntax, but the effect is unchanged.		
The Articles			*Generally:* Compared to JCT 1998 the ncw Articles have been slimmed down with sharper and more contemporary words used.	
Article 1: Contractor's Obligations	Article 1	Slight amendment: the clause repeats the text of JCT 1998 with slight changes that improve the drafting or syntax, but the effect is unchanged.	The Article is a statement that 'the Contractor shall carry out and complete the Works in accordance with the Contract Documents'. The explicit statement that this obligation is undertaken for a consideration has been removed but a valid contract would nevertheless be created as the exchange of obligations is apparent and adequately dealt with by subsequent Articles.	
Article 2: Contract Sum	Article 2	Slight amendment: the clause repeats the text of JCT 1998 with slight changes that improve the drafting or syntax, but the effect is unchanged.		
Article 3: Architect/ Contract Administrator	Article 3	Slight amendment: the clause repeats the text of JCT 1998 but, although it introduces new defined terms and makes slight changes, the effect is unchanged.	Identifies the Architect/Contract Administrator (A/CA), but the Article itself no longer provides processes for the replacement of the consultants. These have been relocated to the main body of the contract (clause 3.5).	

Article 4: Quantity Surveyor	Article 4	Slight amendment: the clause repeats the text of JCT 1998 with slight changes that improve the drafting or syntax, but the effect is unchanged.	Identifies the Quantity Surveyor, but the Article itself no longer provides processes for the replacement of the consultants. These have been relocated to the main body of the contract (clause 3.5).	
Article 5: Planning Supervisor	Article 6.1	Slight amendment: the clause repeats the text of JCT 1998 with slight changes that improve the drafting or syntax, but the effect is unchanged.		
Article 6: Principal Contractor	Article 6.2	Slight amendment: the clause repeats the text of JCT 1998 with slight changes that improve the drafting or syntax, but the effect is unchanged.		
Article 7: Adjudication	Article 5	No change: the clause repeats the text of JCT 1998.		
Article 8: Arbitration	Article 7A	**Major change:** the amendment changes a fundamental process or procedure of the contract compared to JCT 1998.	The most significant change found in the Articles. SBC/Q 2005 does not provide for arbitration unless the Parties specifically select it in the Contract Particulars (arbitration was the default dispute resolution process for JCT 1998). Arbitration can be still be adopted as the default dispute resolution process for the contract by stating this in the Contract Particulars, and SBC/Q 2005 includes clauses that support the arbitration process should it be selected.	The merits of arbitration should be considered, and if it is required the Contract Particulars must be amended.
Article 9: Legal proceedings	Article 7B	Clause redrafted: the clause has been significantly reworded and gives additional clarity, but does not give rise to new obligations.		As above, the merits of using the courts should be considered.

**JCT 2005 Standard Building Contract
With Quantities (SBC/Q 2005)**

SBC/Q 2005 clause and reference title	JCT 1998 clause and reference title	Summary of change	Consequence of change and comments	New action required
Contract Particulars	Appendix		*Generally:* The Contract Particulars provide the project-specific information that was previously found in the Appendix at the end of JCT 1998. The Contract Particulars of SBC/Q 2005 are significantly expanded in comparison to the Appendix of JCT 1998.	
Part 1 General				
Fourth Recital and clause 4.7 (CIS)	Fourth Recital and 31	No change: the clause repeats the text of JCT 1998.		
Sixth Recital (Description of Sections)	Sectional Completion Supplement	**Major change:** the amendment changes a fundamental process or procedure of the contract compared to JCT 1998.	This is a new provision that was not found in the 1998 Sectional Completion Supplement. The descriptions must be shown or the relevant documents referenced. When completing the contract, thought must be given as to the appropriate way to divide the project into Sections.	
Eighth Recital (Employer's Requirements)	CDP Supplement (Supplemental Appendix)	Slight amendment: the clause repeats the text of the 1998 CDP Supplement with slight changes that improve the drafting or syntax, but the effect is unchanged. If required, the option must be selected in the Contract Particulars.	The documents containing the Employer's Requirements are to be referenced or identified here. The slight change is that SBC/Q 2005 contains a note stating the details that identify the Employer's Requirements that was not found in the 1998 CDP Supplement.	Location of the Employer's Requirements in the Contract Documents is to be stated. It perhaps makes sense for the Contract Documents to contain a discrete annex or appendix which, in the interests of clarity, contains solely the Employer's Requirements.

Ninth Recital (Contractor's Proposals)	CDP Supplement (Supplemental Appendix)	Slight amendment: the clause repeats the text of the 1998 CDP Supplement with slight changes that improve the drafting or syntax, but the effect is unchanged. If required, the option must be selected in the Contract Particulars.	The documents containing the Contractor's Proposals are to be referenced or identified here. The slight change is that SBC/Q 2005 contains a note stating the details that identify the Contractor's Proposals that was not found in the 1998 CDP Supplement.	The location of the Contractor's Proposals in the Contract Documents is to be included. It perhaps makes sense for the Contract Documents to contain a discrete annex or appendix, for the avoidance of doubt, which contains solely the Contractor's Proposals.
Ninth Recital (CDP Analysis)	CDP Supplement (Supplemental Appendix)	Slight amendment: the clause repeats the text of the 1998 CDP Supplement with slight changes that improve the drafting or syntax, but the effect is unchanged. If required, the option must be selected in the Contract Particulars.	The documents containing the CDP Analysis are to be referenced or identified here. The slight change is that SBC/Q 2005 contains a note stating the details that identify the CDP Analysis that was not found in the 1998 CDP Supplement.	The location of the CDP Analysis in the Contract Documents is to be included. It perhaps makes sense for the Contract Documents to contain a discrete annex or appendix, for the avoidance of doubt, which contains solely the CDP Analysis.
Article 8	Articles 7A and 7B	**Major change:** the amendment changes a fundamental process or procedure of the contract compared to JCT 1998.	JCT 1998 contained a clause that required that Parties in dispute would take that dispute to an Arbitrator rather than to the courts. In practice, the courts would stay legal proceedings in the courts in deference to the *Arbitration Act* 1996 and the Parties' agreement to adopt arbitration. The new default position for SBC/Q 2005 is diametrically opposite to that stated in JCT 1998 and is that arbitration will not apply to disputes arising under the contract and that disputes will be resolved through the courts.	

SBC/Q 2005 clause and reference title	JCT 1998 clause and reference title	Summary of change	Consequence of change and comments	New action required
			If the Parties want arbitration to be the dispute resolution method for the contract instead of litigation, this must be stated in the Contract Particulars. Failing such agreement for a particular dispute, the courts will not stay legal proceedings in favour of arbitration unless the arbitration has been agreed and initiated by both Parties on an ad hoc basis. JCT 2005 does not interfere with the position that the Parties are free to select arbitration on a case-by-case or ad hoc basis, even though they have not made arbitration the default dispute resolution mechanism. Note also that adjudication will apply to construction contracts irrespective of the selection made.	
1.1 Base Date	1.3 Base Date	No change: the clause repeats the text of JCT 1998.		
1.1 Date for Completion of the Works	1.3 Date for Completion	Slight amendment: the clause repeats the text of JCT 1998 with slight changes that improve the drafting or syntax, but the effect is unchanged.		
1.1 Sections	Sectional Completion Section 1.3 Dates for completion of Sections	Slight amendment: the clause repeats the text of the 1998 Sectional Completion Supplement with slight changes that improve the drafting or syntax, but the effect is unchanged.		

| 1.7 (address for notices) | 1.7 | **Significant change:** the clause repeats most of the text of JCT 1998, but this is amended in a way that significantly affects the operation of the clause. | This new clause gives greater certainty than JCT 1998 as it provides a definite location for correspondence – either the address stated in the Contract Particulars or, in its absence, the address at the beginning of the Articles of Agreement. This avoids any doubts that might arise when it is difficult to determine which of several offices is the Party's principal office and should ensure that a Party needing to issue a notice will always be able to serve a valid notice on the other Party. | |
| 1.8 Electronic communications | 1.11 Electronic Data Interchange | **Significant change:** the clause repeats most of the text of JCT 1998, but this is amended in a way that significantly affects the operation of the clause. | The provision in JCT 1998 for an 'Electronic Data Interchange' (EDI) has been deleted.
The other change of significance is that the Contract Particulars provide a space for the Parties to make a list of the communication, documents and notices etc. that can be validly sent electronically.
Note that if the Parties do not provide such a list, then the contract default position is that communications must be in writing and the consequence would be that electronic documents would have no formal status under the contract.
The change is that the Parties are now free to agree that the listed communications may be validly issued by electronic means. JCT has not suggested which of the communications required by the contract might sensibly be transferred electronically. At one end of the sliding-scale would be a notice of termination, which it is suggested would not be appropriate for electronic notice; and at the other end, the lower-level enquiries and requests for information where it would be practical. | Decide whether or not to consent to valid notice or communication by email, etc.
If this is selected, then the specific items of correspondence and the form of communication must be set down in this section of the Contract Particulars. |

27

SBC/Q 2005 clause and reference title	JCT 1998 clause and reference title	Summary of change	Consequence of change and comments	New action required
			Difficulty would arise with the matters in the middle of the scale; for example, perhaps it is appropriate for withholding notices to be sent by email but inappropriate for instructions to be sent by email.	
2.4 Date of Possession of the Site	Appendix 1 reference to 23.1.1 and Sectional Completion Supplement	Slight amendment: the clause repeats the text of JCT 1998 with slight changes that improve the drafting or syntax, but the effect is unchanged.	Note that if the site is to be provided in Sections, then these Sections and dates are to be stated here.	
2.5 and 2.29.3 Deferment of possession	23.1.2 Deferment of possession	Slight amendment: the clause repeats the text of JCT 1998 with slight changes that improve the drafting or syntax, but the effect is unchanged.	Note that the Parties may opt to provide different periods of maximum deferment to particular Sections if the work is to be carried out in Sections.	
2.19.3 CDP Limit of Contractor's liability	CDP Supplement: Supplementary Appendix	Slight amendment: the clause repeats the text of JCT 1998 with slight changes that improve the drafting or syntax, but the effect is unchanged.	There is a small degree of change in that this section of the Contract Particulars states that the liability need not be capped at all. If no figure is stated in the Contract Particulars, then the liability of the Contractor would be uncapped.	
2.32.2 Liquidated Damages	Appendix and Appendix of Sectional Completion Supplement	Slight amendment: the clause repeats the text of JCT 1998 with updated definitions and cross-references for SBC/Q 2005, but the effect is unchanged.	SBC/Q 2005 refers to 'liquidated damages' rather than 'liquidated and ascertained damages' referred to in JCT 1998 and the 1998 Sectional Completion Supplement. Otherwise the clause is unchanged.	The user should always take great care as to the calculation and negotiation of the amount included here.

2.37 Sections: Sums	Appendix Sectional Completion Supplement 18.1.5 Section value	Slight amendment: the clause repeats the text of JCT 1998 with updated definitions and cross-references for SBC/Q 2005, but the effect is unchanged.	
2.38 Rectification Period(s)	17.2 Defects Liability Period	Slight amendment: the clause repeats the text of JCT 1998 with updated definitions and cross-references for SBC/Q 2005, but the effect is unchanged.	The Defects Liability Period has been re-branded as a 'Rectification Period'. There is no consequence as a result of the change, but perhaps 'snags' are more obviously covered by the definition.
4.8 Advance Payment	30.1.1.6 Advance Payment	No change: the clause repeats the text of JCT 1998.	Note that this provision does not apply if the Contractor is a local authority.
4.8 Advance Payment Bond	30.1.1.6 Advance Payment	**Major change:** the amendment changes a fundamental process or procedure of the contract compared to JCT 1998.	An Advance Payment Bond must be automatically provided by the Contractor in connection with an advance payment as a default unless the Contract Particulars state here that this is not required. JCT 1998 made the provision of an Advance Payment Bond optional rather than mandatory. Advance Payments do not apply if the Employer is a local authority.
4.9.2 Date of issue of Interim Certificates	30.1.3	Slight amendment: the clause repeats the text of JCT 1998 with slight changes that improve the drafting or syntax, but the effect is unchanged.	
4.17.4 Listed Items	30.1.1	Slight amendment: the clause repeats the text of JCT 1998 with slight changes that improve the drafting or syntax, but the effect is unchanged.	An additional guidance note in the clause avoids potential confusion by stating that the provision does not apply if a 'Bond in respect of payment for off-site materials and/or goods' is not required (Schedule 6 Part 2).

SBC/Q 2005 clause and reference title	JCT 1998 clause and reference title	Summary of change	Consequence of change and comments	New action required
4.17.5	30.3.2	Slight amendment: the clause repeats the text of JCT 1998 with slight changes that improve the drafting or syntax, but the effect is unchanged.	An additional guidance note in the clause avoids potential confusion by stating that the provision does not apply if a 'Bond in respect of payment for off-site materials and/or goods' is not required (Schedule 6 Part 2).	
4.19 Contractor's Retention Bond	30.4A	Slight amendment: the clause repeats the text of JCT 1998 with slight changes that improve the drafting or syntax, but the effect is unchanged.	The default position is that this bond is not required unless the Contract Particulars state here that it is to be provided. Note that this provision does not apply if the Contractor is a local authority.	
4.20.1 Retention Percentage	Retention Percentage	**Major change:** the rights of the Parties have been changed by this amendment compared to JCT 1998.	The default Retention Percentage is 3% for SBC/Q 2005, whereas JCT 1998 provided for a rate of 5%.	The percentage sum is different as a result of this change, and thus any standard documentation should be adjusted to reflect this change.
4.21 and Schedule 7 Fluctuations Options	Appendix	**Major change:** the amendment changes a fundamental process or procedure of the contract compared to JCT 1998.	The Fluctuations Options are more thoroughly set out in SBC/Q 2005 than in JCT 1998. The standard published version of JCT 1998 did not include the text for the three Fluctuations Options (described as clauses 38, 39 and 40) – these are now included in SBC/Q 2005 as Schedule 7. SBC/Q is unchanged in that if no option is selected the Contract defaults to Option A (clause 38 as was). The Contract Particulars default to Part I of the Formula Rules in the absence of any other entry in the Contract Particulars.	If the Parties require a more absolute lump sum contract, then consideration should be given to the deletion of all of the fluctuation clauses.

6.4.1 Contractor's insurance – injury to persons or property	21.1.1	Clause redrafted: the clause has been significantly reworded and gives additional clarity, but does not give rise to new obligations.		
6.5.1 Insurance – Liability of the Employer	21.2.1 Insurance Liability of the Employer	Clause redrafted: the clause has been significantly reworded and gives additional clarity, but does not give rise to new obligations.	The new note reinforces the default position that the insurance is not required in the absence of a specific opt-in evidenced by the completion of this part of the Contract Particulars.	
6.7 and Schedule 3 Insurance of the Works	22.1 Insurance of the Works	Slight amendment: the clause repeats the text of JCT 1998 with cross-references and definitions for SBC/Q 2005, but the effect is unchanged.		To avoid confusion users should ensure that only one option is selected.
6.7 and Schedule 3 Insurance Option A	22A Percentage to cover professional fees	**Major change:** the amendment changes a fundamental process or procedure of the contract compared to JCT 1998.	The provision of a default position in the absence of any preference being stated is a major change. This was not provided in JCT 1998. SBC/Q 2005 provides that a coverage of 15% is the default sum required in respect of professional fees connected with rectification of insured risk events.	Users who require cover for professional fees should check that the default percentage is correct for their particular project.
6.7 and Schedule 3 Insurance Option A	22A Annual renewal date	Slight amendment: the clause repeats the text of JCT 1998 with updated definitions and cross-references for SBC/Q 2005, but the effect is unchanged.		
6.11 Contractor's Designed Portion Professional Indemnity Insurance		**Major change:** the amendment changes a fundamental process or procedure of the contract compared to JCT 1998 and the CDP Supplement.	This optional term makes provision for the Contractor to insure his design liability. This was neither a requirement nor an option of JCT 1998 nor was it a provision of the 1998 CDP Supplement. Note that if no sum is stated, then the Contractor will not be obliged by the contract to insure his liability for his CDP.	State the value of professional indemnity insurance that the Contractor must carry in respect of the CDP

31

JCT 2005 Standard Building Contract
With Quantities (SBC/Q 2005)

SBC/Q 2005 clause and reference title	JCT 1998 clause and reference title	Summary of change	Consequence of change and comments	New action required
6.13 Joint Fire Code	22FC.1 Joint Fire Code	Slight amendment: the clause repeats the text of JCT 1998 with updated definitions and cross-references for SBC/Q 2005, but the effect is unchanged.		
6.16 Joint Fire Code	22FC.5 Joint Fire Code	**Significant amendment:** the clause repeats most of the text of JCT 1998, but this is amended in a way that does effect the operation of the clause.	If the Contract Particulars do not state a preference, then the default position will apply and the Contractor will be obliged to bear the cost of compliance with amendments or revision to the Joint Fire Code.	
7.2 Assignment by Employer	19.1.2	Slight amendment: the clause repeats the text of JCT 1998 with updated definitions and cross-references for SBC/Q 2005, but the effect is unchanged.		
8.9.2 Period of suspension	28.2.2	Slight amendment: the clause repeats the text of JCT 1998 with updated definitions and cross-references for SBC/Q 2005, but the effect is unchanged.	The time-period for suspension has been increased from one month to two months in SBC/Q 2005.	
8.11.1.1 to 8.11.1.5 Period of suspension	28A.1.1.1–6	**Significant change:** the clause repeats most of the text of JCT 1998, but this is amended in a way that significantly affects the operation of the clause.	The time-period for suspension has been standardised at two months in SBC/Q 2005. JCT 1998 provided that the period for clauses 28A.1.1.1–3 was to be three months, and for 28A.1.1.4–6 was to be one month.	
9.2.1 Adjudication	41A.2	**Major change:** the amendment changes a fundamental process or procedure of the contract compared to JCT 1998.	Adjudicator appointment under SBC/Q 2005 is different from JCT 1998 in a number of respects.	

			SBC/Q 2005 provides that the Parties may nominate a specific individual in the Contract Particulars to act as Adjudicator – this was not provided for in JCT 1998. If the Contract Particulars have not been completed, then a referring Party may select any of the listed Adjudicator Nominating Bodies (ANBs). Under JCT 1998 the contract defaulted to the RIBA as ANB.	
9.4.1 Arbitration	41B.1	Slight amendment: the clause repeats the text of JCT 1998 with slight changes that improve the drafting or syntax, but the effect is unchanged.	The remit of the appointer under SBC/Q 2005 is slightly wider (or at least the role is more explicit) than under JCT 1998 in that the appointment of a replacement Arbitrator is also covered. As such, this is a more complete draft that provides a resolution when a specific difficulty is encountered.	
Part 2 Third Party Rights and Collateral Warranties	None	**Major change:** this is a wholly new clause for the SBC/Q 2005 and the balance of risk is changed as a result.	The third party rights and collateral warranty regime for SBC/Q 2005 is new and has no precedent in JCT 1998. It is a series of options (set out in full in Section 7) that provides that the Contractor (and in some instances the sub contractor) accepts certain obligations to persons other than the Employer. In the past, Parties have tended to draft their own pro forma warranties or adopt the separate JCT Collateral Warranty Supplement. They have also tended to exclude the application of the *Contracts (Rights of Third Parties) Act* 1999. The actual rights themselves are set out in clause 7 and Schedule 5 of the Contract, but it is crucial that if these rights are required that this is indicated at this point in the Contract Particulars.	Users must decide if they want to provide third party rights and who should benefit from them. Some clients or Funders may wish to continue to rely upon their own standard warranties. Unless there is a specific statement as to the rights that are to apply, then there will be no third party rights or collateral warranties under the contract.

SBC/Q 2005 clause and reference title	JCT 1998 clause and reference title	Summary of change	Consequence of change and comments	New action required
			The default position is that there are to be no rights under the contract for Purchasers/Tenants or Funders. Users should consider, understand, negotiate and clearly record the identity of those third parties that they wish to grant rights, as well as noting those rights that they are willing to allow. For all users, issues such as levels of insurance, limits on liability, duties of care and assignment now become issues that require earlier consideration in the progress towards execution of the contract.	
Attestation	Attestation page headed 'as witness the hands of the parties hereto'	Clause redrafted: the clause has been significantly reworded and gives additional clarity, but does not give rise to new obligations.	The attestation process is unchanged, but the words used are more modern and the notes in SBC/Q 2005 itself are clearer than those in JCT 1998. Like JCT 1998, SBC/Q 2005 provides attestation forms for limited companies and individuals. The guidance note points out that the attestation forms may not be correct for housing associations, partnerships and foreign companies, but this is not stated in the terms of the contract itself.	Confirm that the method of execution is correct for the Party entering the Agreement.
CONDITIONS				
Section 1: Definitions and Interpretation			Several definitions that had been found in the text of JCT 1998 are now located, or at least referenced, in this part of SBC/Q 2005.	

			The definitions also serve as something approaching an index, as if the term is defined in the text of the main contract this is noted and the clause reference is given. Definitions that are not amended, or where the change does not have material effect, are not noted here.	
Definitions				
Activity Schedule	Activity Schedule	**Significant amendment:** the clause has been significantly reworded and gives additional clarity, but does not give rise to new obligations.	The definition in SBC/Q 2005 is much shorter than the JCT 1998 clause. The old clause was more prescriptive in terms of what had to be stated for the Activity Schedule to be 'fully priced'.	
Architect/Contract Administrator	Architect	**Significant amendment:** the clause has been significantly reworded, but does not give rise to new obligations.	SBC/Q 2005 adopts the practice of other JCT contracts (but not JCT 1998) and has adopted the phrase 'Architect/Contract Administrator' (A/CA) in place of 'Architect'. Users may read great significance into this, but the processes of the contract are not affected. The selection of the correct Contract Administrator for the project will remain one of the most significant decisions in the project lifecycle that an Employer makes.	
Business Day	N/A	**New definition:** this clause provides a new definition of a term or of a process.	The definition 'Business Day' is new, however the reckoning of periods of days (clause 1.5) and definition of Public Holiday are unchanged. This is relevant because many of the time-periods stated in the contract make reference to 'days' not 'Business Days' (e.g. clause 3.12, the confirmation of verbal instructions, makes no reference to Business Days), which means that users	Note which units of time are to be applied to each time-period stated in the contract.

SBC/Q 2005 clause and reference title	JCT 1998 clause and reference title	Summary of change	Consequence of change and comments	New action required
			should note which units of time they are dealing with as there are usually consequences for failing to get it right.	
CDP Analysis CDP Documents CDP Works	Contractor's Designed Portion Supplement	Slight amendment: the clause repeats the text of the 1998 CDP Supplement with slight changes that improve the drafting or syntax, but the effect is unchanged.		
Certificate of Making Good	Certificate of Completion of Making Good Defects	Clause redrafted: the clause has been significantly reworded, but does not give rise to new obligations.	The name of the certificate has changed, however its function is unchanged. The definition itself is stated in SBC/Q 2005 at clause 2.39.	If you make use of standard letters or notices, then these must be updated to reflect the new terminology and definitions used.
Contract Documents	Contract Documents	Slight amendment: the clause repeats the text of the 1998 CDP Supplement with slight changes that improve the drafting or syntax, but the effect is unchanged.	There is a minor expansion of the scope of the term to permit inclusion of documents relating to the CDP and also to recognise that certain material that was in the Appendix is now in the Contract Particulars. If there is to be no CDP there is no material change.	
Contract Particulars	Appendix	**New definition:** this clause provides a new definition of a term or of a process.	This is a new definition describing the Contract Particulars themselves and also noting that they are completed by the Parties.	
Contractor's Design Documents	Contractor's Designed Portion Supplement	**New definition:** this clause provides a new definition of a term or of a process.		

Contractor's Designed Portion Contractor's Proposals	Contractor's Designed Portion Supplement	Slight amendment: the clause repeats the text of the 1998 CDP Supplement with slight changes that improve the drafting or syntax, but the effect is unchanged.		
Contractor's Persons	N/A	**New definition:** this clause provides a new definition of a term or of a process.	Useful catch-all definition that describes who the Contractor is directly responsible for.	
Employer's Persons	N/A	**New definition:** this clause provides a new definition of a term or of a process.	Catch-all definition that describes who the Employer is directly responsible for. Note that the A/CA and the Quantity Surveyor are expressly excluded from the scope of this definition.	
Employer's Requirements	Contractor's Designed Portion Supplement Additional Definition	Slight amendment: the clause repeats the text of the 1998 CDP Supplement with slight changes that improve the drafting or syntax, but the effect is unchanged.	New definition to facilitate the CDP that is contained in the Eighth Recital.	
Finance Agreement Funder Funder Rights Particulars	N/A	**Major change:** the amendment changes a fundamental process or procedure of the contract compared to JCT 1998.	Four new definitions to accommodate the provision of third party rights or warranties for Funders.	
Health and Safety Plan	Health and Safety Plan	**Significant amendment:** the clause repeats most of the text of JCT 1998, but this is amended in a way that does effect the operation of the clause.	The second half of the JCT 1998 definition has been removed to clause 3.25 'CDM Regulations'.	
P&T Rights P&T Particulars Purchaser	N/A	**Major change:** the rights of the Parties have been changed by this amendment compared to JCT 1998.	These three new definitions support the new regime for providing rights for Purchasers and Tenants.	
Rectification Period	Defects Liability Period	Clause redrafted: the clause has been significantly reworded, but does not give rise to new obligations.	This clause brings a change of the name of the period for correcting defects.	
Retention Percentage	Retention Percentage	**Major change:** the rights of the Parties have been changed by this amendment compared to JCT 1998.	Users familiar with JCT 1998 should note that the Retention Percentage is 3% for SBC/Q 2005. The JCT 1998 provided for retention at a rate of 5%.	Adjust standard processes to retain 3% rather than 5%. Valuation and payment forms should be

37

**JCT 2005 Standard Building Contract
With Quantities (SBC/Q 2005)**

SBC/Q 2005 clause and reference title	JCT 1998 clause and reference title	Summary of change	Consequence of change and comments	New action required
				adjusted to reflect the new default percentage. Also, tenders should reflect the change in cashflow throughout the duration of the Works.
Interpretation				
1.2 Reference to Clauses etc.	1.1, 2.2.1	Clause redrafted: the clause has been significantly reworded and gives additional clarity, but does not give rise to new obligations.		
1.3 Articles etc. to be read as a whole	1.2	Clause redrafted: the clause has been significantly reworded (and gives additional clarity), but does not give rise to new obligations.	Note that the Contract Particulars form part of the Articles of Agreement.	
1.4 Headings, reference to persons, legislation etc.	N/A	**Significant amendment:** the clause has been significantly reworded, but does not give rise to new obligations.	This clause might be termed a collection of legal 'boiler plate' that ensures that words are not taken out of context – it does not of itself create any new obligations but is of value in supporting the operation of the contract.	
1.5 Reckoning periods of days	1.8	Slight amendment: the clause repeats the text of JCT 1998 with slight changes to the text, but the effect is unchanged.		
1.6 Contracts (Rights of Third Parties) Act 1999	1.12	**Major change:** this is a wholly new clause for the SBC/Q 2005 and the balance of risk is changed as a result.	The Act may be used as part of the scheme to provide rights for Purchasers and Tenants and Funders, but is otherwise excluded unless it has been specifically adopted in this respect. This is a fundamental change that is explored more fully in Section 7 under third party rights.	

1.7 Giving or service of notices and other documents	1.7	Clause redrafted: the clause has been significantly reworded and gives additional clarity, but does not give rise to new obligations.	The rewording has added the phrase 'actual delivery' to describe the process of hand delivery of a notice.	
1.8 Electronic Communications	1.11	**Major change:** the amendment changes a fundamental process or procedure of the contract compared to JCT 1998.	The provision for an Electronic Data Interchange found in JCT 1998 has been deleted. The Parties may state that email and other forms of electronic communication will be a valid form of communication and may proscribe its validity for certain kinds of communication in accordance with their preference. Adoption of this change is certainly not mandatory – indeed, to do so may be considered wholly inappropriate for some projects. The Contract Particulars provide a space for the Parties to make a list of the documents that can be validly sent electronically. JCT has not suggested which of the communications required by the contract might sensibly be transferred electronically. At one end of the sliding-scale would be a notice of termination, which it is suggested would not be appropriate for electronic notice; and at the other end, the lower-level enquiries and requests for information, where it would be practical.	Complete the Contract Particulars if electronic communication is considered satisfactory or appropriate for key contract processes.
1.9 Issue of Architect/ Contract Administrator's certificates	5.8	Slight amendment: the clause repeats the text of JCT 1998 with slight changes that improve the drafting or syntax, but the effect is unchanged.		

39

SBC/Q 2005 clause and reference title	JCT 1998 clause and reference title	Summary of change	Consequence of change and comments	New action required
1.10 Effect of Final Certificate	30.9	Slight amendment: the clause repeats the text of JCT 1998 with slight changes that improve the drafting or syntax, but the effect is unchanged.	Matters concerning the finality of the Final Certificate in the context of the adjudication and arbitration found in JCT 1998 under clauses 30.9.2 and 30.9.3 are now merged in to a single clause 1.10.2. An example of the better use of defined terms and improved drafting in SBC/Q 2005 is found here: the contract now refers to 'Relevant Matters' at 1.10.4 where it had previously simply referred to 'matters referred to in clause 26.2' – this is both shorter and clearer.	
1.11 Effect of certificates other than Final Certificate	30.10	Slight amendment: the clause repeats the text of JCT 1998 with slight changes that improve the drafting or syntax, but the effect is unchanged.		
1.12 Applicable Law	1.10	Clause redrafted: the clause has been significantly reworded, but does not give rise to new obligations.	Fewer words are used to ensure the same provision.	
Section 2: Carrying out the Works				
Contractor's Obligations				
2.1 General Obligations	2.1, 6.1.1	Clause redrafted: the clause has been significantly reworded (and gives additional clarity), but does not give rise to new obligations.	The Contractor's general obligation stated in this clause is a stripped-down version of the previous clause. There is now an express obligation to carry out the Works in compliance with the Health and Safety Plan and the Statutory Requirements. The clause obliges the Contractor to comply with the Contract Documents. Accordingly, if the standard for the Works	

			is the reasonable satisfaction of the Architect/Contract Administrator or some objective standard, this must be stated in the Contract Documents. The definition of 'Works' includes the Works required by the Contractor's Designed Portion, and thus this clause also states the standards for the CDP Works too.	
2.2 Contractor's Designed Portion	Contractor's Designed Portion Supplement Amendments to clauses 2.1–2 of JCT 1998	Slight amendment: the clause repeats the text of the 1998 CDP/Sectional Completion Supplement with slight changes that improve the drafting or syntax, but the effect is unchanged.		It is very important for the Employer to state a standard of workmanship for the CDP Works in the Employer's Requirements.
2.3 Materials, goods and workmanship	2.3.3, 8.1.1, 8.1.5, 8.2.1 and Contractor's Designed Portion Supplement	Clause redrafted: the clause has been significantly reworded and gives additional clarity, but does not give rise to new obligations.	Clause 2.3 is a sandwich of two familiar clauses from JCT 98 (the old clauses 8.1.1 and 8.1.5) with new wording to accommodate the new facility for a CDP. The standard stated is familiar – i.e. compliance with the Contract Bills so far as procurable and a prohibition on substitution of materials. Clause 2.3.2 establishes that the Contract Documents will describe the standard of workmanship for the CDP. It should be noted that if the Employer creates a CDP but fails to state a standard of workmanship in the Employer's Requirements, then the standard of workmanship that the Contractor has stated in his Contractor's Proposals will apply. Clause 2.3.4 obliges the Contractor to prove his compliance with the standards for materials and goods. The JCT 1998 contract obliged the Contractor to provide 'vouchers' to this effect, whereas the present version requires 'reasonable proof' – this requirement would appear to be more contemporary if less specific.	Employer to state the standard of workmanship for CDP in the Employer's Requirements.

SBC/Q 2005 clause and reference title	JCT 1998 clause and reference title	Summary of change	Consequence of change and comments	New action required
Possession				
2.4 Date of possession – progress	23.1, 23.3.1	Clause redrafted: the clause has been significantly reworded and gives additional clarity, but does not give rise to new obligations.	The clause groups obligations more effectively and accommodates the work being carried out in Sections, but the balance of risk remains the same as in JCT 1998.	
2.5 Deferment of possession	23.1.2	Clause redrafted: the clause has been significantly reworded and gives additional clarity, but does not give rise to new obligations.	If the Employer wishes or needs to retain the right to delay the possession, then it must state this in the Contract Particulars.	
2.6 Early use by the Employer and others	23.3.2, 23.3.3	Slight amendment: the clause repeats the text of JCT 1998 with slight changes that improve the drafting or syntax, but the effect is unchanged.		
2.7 Work not forming part of the Contract	29	**Significant change:** a new clause that enables a new process or action from the Parties.	The Employer is not obliged to give all of its work to the Contractor and this clause provides that work that might be on the site or intrinsically related to it (but not included in the contract) may be carried out by the Employer (or perhaps his Purchaser's fit out contractor). If this is anticipated and can be worked into the Contract Bills, then so much the better as the Contractor must permit access for the workmen. Otherwise the consent of the Contractor will be necessary.	Employer should list, where possible, specifically or generically, workmen that will require access whilst the site is in the possession of the Contractor.
Supply of Documents, Setting Out etc.				
2.8 Contract Documents	5.1, 5.2, 5.4, 5.5, 5.7	**Significant change:** the clause repeats much of the text of JCT 1998, but also provides for new process or action from the Parties.	The Contract Documents are to be available for inspection by the Contractor and SBC/Q 2005 requires that these will be retained by the Employer.	Employer to retain documents identified and to be prepared to make them available to the Contractor.

			JCT 1998 required the same documents to be available for inspection, but required that the Architect or Quantity Surveyor retain them rather than the Employer. 2.8.4 safeguards commercially sensitive information and is partially a confidentiality and intellectual property rights clause, it seeks to prevent disclosure of the Contractor's rates and (mis)appropriation of the design documents and is of similar effect to clause 5.4 of JCT 1998.
2.9 Construction Information and Contractor's master programme	5.3.1, 5.3.2	Slight amendment: the clause repeats the text of JCT 1998 with slight changes that improve the drafting or syntax, but the effect is unchanged.	The programme requirements are stated here and are substantially the same as JCT 1998. The master programme remains a tool for checking progress rather than a method of creating obligations. There is no provision whereby the Contractor might be considered in 'breach of master programme'. At 2.9.1 the clause provides for the documentation concerning the CDP to be provided to the Architect/Contract Administrator. 2.9.3 is an additional clause that incorporates the Contractor's Design Submission Procedure in Schedule 1 of SBC/Q 2005. This process was not found in JCT 1998, but it is similar to the procedure first seen in the JCT Major Projects Form of 2003. The Contractor's Designed Submission Procedure is considered in the notes against Schedule 1 of SBC/Q.

43

SBC/Q 2005 clause and reference title	JCT 1998 clause and reference title	Summary of change	Consequence of change and comments	New action required
2.10 Levels and setting out of the Works	7	Slight amendment: the clause repeats the text of JCT 1998, but this is amended to accommodate the incorporation of the CDP provisions into SBC/Q 2005.	Obligation that is substantially unchanged compared to JCT 1998 save that it now makes provision for the possibility of a Contractor's Designed Portion.	
2.11 Information Release Schedule	5.4	Slight amendment: the clause repeats the text of JCT 1998 with slight changes that improve the drafting or syntax, but the effect is unchanged.		
2.12 Further drawings, details and instructions	5.4.2	Slight amendment: the clause repeats the text of JCT 1998 with slight changes that improve the drafting or syntax, but the effect is unchanged.	Compared to clause 5.4.2 of JCT 1998 this is a much improved statement (to the extent that it is now comprehensible) of what documentation is required from which Party and at which point.	
Errors, Discrepancies and Divergences				
2.13 Preparation of Contract Bills and Employer's Requirements	2.2.2	Clause redrafted: the clause has been significantly reworded (and gives additional clarity), but does not give rise to new obligations.	The latest version of the Standard Method of Measurement (7th Edition) is to be used to prepare the Contract Bills: this does not represent any change. However, clause 2.13.2 is of particular significance: '... the Contractor shall not be responsible for the contents of the Employer's Requirements or for verifying the accuracy of any design contained within them'. This change is a clarification of the existing balance of risk rather than a change to the balance of risk. The SBC/Q retains the traditional position of design	

			responsibility and design liability being retained by the Employer. Under this form of contract the Employer has ultimate recourse to the Architect on the issue of design liability. The clause has also made provision for the specific exception to this rule where the Contractor accepts the design responsibility for the CDP.	
2.14 Contract Bills and CDP Documents – errors and inadequacies	2.2.2.2	Clause redrafted: the clause has been significantly reworded and gives additional clarity but does not give rise to new obligations.	If there is an error in the Contract Bills then, as with JCT 1998, there is to be a Variation. The Guidance Note underlines the point that 'the Contractor should not necessarily be required to check all preliminary design work of the Employer's advisers'. Under clause 2.14.2, if there is an error in the Employer's Requirements and the Contractor corrects the Employer's error in his Proposals, and hence before the Contract has been entered into, then his solution applies – if the Contractor has not corrected the error, then there is to be a Variation.	
2.15 Notification of discrepancies	2.3.1–2.3.5	**Significant change:** the clause repeats most of the text of JCT 1998 but this is amended in a way that significantly affects the operation of the clause.	The provision in SBC/Q 2005 is less prescriptive than it was in JCT 1998, as in that form the Contractor had to specify the discrepancy or divergence. SBC/Q 2005 requires the Contractor to provide appropriate details thus a Contractor must exercise an element of judgement as to what is required.	

45

JCT 2005 Standard Building Contract
With Quantities (SBC/Q 2005)

SBC/Q 2005 clause and reference title	JCT 1998 clause and reference title	Summary of change	Consequence of change and comments	New action required
2.16 Discrepancies in CDP Documents	Contractor's Designed Portion Supplement Clauses 2.5.1.1–2	Slight amendment: the clause repeats the text of the 1998 CDP Supplement with slight changes that improve the drafting or syntax, but the effect is unchanged.	If the Contractor is culpable, then the costs of dealing with that discrepancy are not added to the Contract Sum (2.16.1). If the culpability for the discrepancy belongs to the Employer, and the Contractor has resolved the discrepancy in his Proposals, then this clause will apply. Otherwise the Contractor and A/CA are to agree what is to be done as a Variation; if they can't agree the A/CA can decide the form of the Variation.	
2.17.1 Divergences from Statutory Requirements	6.1.2, 6.1.3, 6.1.5	Clause redrafted: the clause has been significantly reworded and gives additional clarity, but does not give rise to new obligations.	The clause is a better structured version of the original clause 6.1 of the JCT 1998.	
2.18 Emergency compliance with Statutory Requirements	6.1.4	Slight amendment: the clause repeats the text of JCT 1998 with slight changes that improve the drafting or syntax, but the effect is unchanged.	If an emergency arises, Contractors should keep reasonably detailed lists of the materials used and the man-hours expended in managing the emergency. Employers and their representatives should then be prepared to assess whether the Contractor is responsible for the emergency, and if not a Variation should be agreed.	
CDP Design Work				
2.19 Design liabilities and limitation	Contractor's Designed Portion Supplement clauses 2.7.1–3	Slight amendment: the clause repeats the text of the 1998 CDP Supplement with slight changes that improve the drafting or syntax, but the effect is unchanged.	This clause provides that a Contractor providing design services owes no greater duty than an Architect would. In addition, the clause also retains the 1998 CDP Supplement provision for a cap on liability. The underlying point is that the Contractor does not acquire a fitness-for-purpose obligation.	

2.20 Errors and failures – other consequences	Contractor's Designed Portion Supplement clauses 2.10.1–3	Slight amendment: the clause repeats the text of the 1998 CDP Supplement with slight changes that improve the drafting or syntax, but the effect is unchanged.	The CDP Supplement provisions concerning errors and failures in the CDP are redrafted, but essentially repeated. The Contractor is not to be given an extension of time if it makes a mistake when compiling the Contractor's Proposals, nor does it get an extension if it submits documents late (including programming documents).
Fees, Royalties and Patent Rights			
2.21 Fees or charges legally demandable	6.2	Significant change: the clause repeats much of the text of JCT 1998, but also provides for new process or action from the Parties.	If charges are due to a statutory body or local authority, then the Contractor pays these. The payment of the fee is expressed as an indemnity (which in this context means the Contractor must pay it), but the sum will generally be added to the Contract Sum. There are two exceptions: (1) The clause provides that if the payment of the fee relates to a Provisional Sum, then it will be dealt with as a Variation and the payment as a disbursement. (2) If the fee arises out of the CDP then the Contractor is assumed to have taken account of it (and thus there is no addition to the Contract Sum). The reference linking Provisional Sums to fees is a development on JCT 1998 as (obviously) is the reference to the CDP. The clause is more concise than the 1998 form and makes reference to the term 'Statutory Requirements' found in the definitions.

47

SBC/Q 2005 clause and reference title	JCT 1998 clause and reference title	Summary of change	Consequence of change and comments	New action required
2.22 Royalties and patent rights – Contractor's indemnity	9.1	Slight amendment: the clause repeats the text of JCT 1998 with slight changes that improve the drafting or syntax, but the effect is unchanged.		
2.23 Patent rights – instructions	9.2	Slight amendment: the clause repeats the text of JCT 1998 with slight changes that improve the drafting or syntax, but the effect is unchanged.		
Unfixed Materials and Goods – property, risk, etc.				
2.24 Materials and goods – on site	16.1	Slight amendment: the clause repeats the text of JCT 1998 with slight changes that improve the drafting or syntax, but the effect is unchanged.		
2.25 Materials and goods – off site	16.2	Slight amendment: the clause repeats the text of JCT 1998 with slight changes that improve the drafting or syntax, but the effect is unchanged.	This clause deals with listed off-site materials/goods. However, if the Employer agrees to pay for off-site materials that are not listed, then it should insist that the materials should be deemed to be listed for the purposes of this clause, in order that the Employer may gain the protection offered by this clause.	
2.26–2.29 Adjustment of Completion Date	Clause 25	Slight amendment: the text of JCT 1998 is repeated, but new defined terms are used, slight changes are made and the clause is amended to accommodate the Sectional Completion and CDP provisions.	The JCT 1998 phrase 'extension of time' has been replaced by 'Adjustment of Completion Date'. This reflects the central idea that the period of time for completion might be reduced as well as being 'extended', for example in the instance where work is no longer to be carried out by the Contractor.	Both Variations and instructions issued by the Architect/Contract Administrator are included as Relevant Events for the adjustment of the Completion Date.

			The provisions remain similar to JCT 1998 and the clause has been improved by the redraft both in terms of layout, substance and comprehensibility. Somewhat confusingly, the phrase 'extension of time' is still found at various points in SBC/Q 2005 but this does not of itself appear to be of consequence. Additions to accommodate working in Sections and provision of a CDP are found throughout the clause. Clause 2.26.2 encourages the Parties to agree any time implications before the Works are varied or instructions are issued.	
2.26 Related definitions and interpretation	25.1	**New definition:** this clause provides a new definition of a term or of a process.	Note the new definition of 'Relevant Omissions' in regard to Provisional Sums that might be instructed.	
2.27 Notice by the Contractor of delay to progress	25.2.1.1–2, 25.2.3	Slight amendment: the clause repeats the text of JCT 1998 with slight changes that improve the drafting or syntax, but the effect is unchanged.	Clause 2.27.1 places a largely unaltered obligation upon the Contractor to give notice of actual or likely delay. The clause sets out the information required as to the causes of the delay. It appears that this obligation maintains the principle that the Contractor does not have to advise of prospective delay. Clause 2.27.3 states an obligation on the Contractor to tell the A/CA of material change to the circumstances forthwith and provides that the A/CA may request information. It is marginally less conditional than its predecessor.	

SBC/Q 2005 clause and reference title	JCT 1998 clause and reference title	Summary of change	Consequence of change and comments	New action required
2.28 Fixing Completion Date 2.28.1	25.3.1, 25.3.2, 25.3.3	**Significant change:** the clause repeats most of the text of JCT 1998, but this is amended in a way that significantly affects the operation of the clause.	The Architect's duty to adjust the date for completion was previously stated to be conditional upon the Contractor having provided notice, particulars and an estimate of the delay. It appears that under SBC/Q 2005 an estimate is not required before the obligation becomes effective upon the A/CA. This is material as the definition of 'notice' for this clause does refer to the provision of an estimate but this is couched in terms of its practicability (clause 2.27.3) and thus the obligation on the A/CA to act may arise before the obligation on the Contractor to provide the estimate. This is consistent with the view that an A/CA must administrate the contract and make decisions rather than issue innumerable requests for estimates and further information. The clause also notes that certain other clauses may specifically remove the A/CA's authority to adjust the time for completion.	The A/CA probably needs to monitor progress more closely than he was required to under JCT 1998 so as to be in a position to make an informed decision where useful information from the Contractor is sparse.
2.28.2	25.3.1	**Significant change:** the clause repeats most of the text of JCT 1998, but this is amended in a way that significantly affects the operation of the clause.	The 2005 form recognises that the A/CA makes a decision whether or not to adjust the time for completion. The word 'decision' is perhaps a more personal term indicating the need for action that suggests the individual A/CA must take a proactive role. The time-period for the A/CA to advise of his decision remains 12 weeks. The period by which the date is adjusted also remains what is 'fair and reasonable' at that time.	

			SBC/Q 2005 states a requirement for the A/CA to 'endeavour' to give the decision in fewer than 12 weeks when the project is fewer than 12 weeks from completion. This appears less absolute than the 1998 form of contract where it was stated that the date was to be fixed 'not later than the Completion Date'. In consequence, the A/CA must make the decision, but is now given the time to make it, despite the approach of the Completion Date. This will remove the pressure to rush out an ill considered certificate. This should assist the Architect/Contract Administrator – particularly given the requirement to provide reasons against adjustments set out under 2.28.3.	
2.28.3	25.3.1.3–4	**Major change:** the amendment changes a fundamental process or procedure of the contract compared to JCT 1998.	Under JCT 1998, where the A/CA was uncertain as to all of the causes of delay, or where it was difficult to state the true cause of the delay, then notices granting revisions to the Completion Date could be lacking in detail. The ability to provide such imprecise revisions to the Completion Date has been reduced by the SBC/Q 2005 wording of the clause. The new clause 2.28.1.3.1 provides that the A/CA shall state the extension of time attributed to each Relevant Event and not just the list of Relevant Events that he has taken into account. This suggests period of time as a number of days or weeks to be stated against each Relevant Event. It is suggested that if a matter is a Relevant Event but no time is due against it, that the A/CA must state that a period of zero weeks be stated. In order to provide a complete response, it is suggested that it	A/CA is to provide a response to each notice for an adjustment to the Completion Date from the Contractor. The A/CA's response is to state the period of time he has awarded against each Relevant Event.

51

JCT 2005 Standard Building Contract With Quantities (SBC/Q 2005)

SBC/Q 2005 clause and reference title	JCT 1998 clause and reference title	Summary of change	Consequence of change and comments	New action required
			would be necessary for the A/CA to state that the matter the Contractor gave notice of was not a Relevant Matter. The A/CA therefore has to provide a detailed report of his decisions stating: • which of the matters notified are Relevant Events; • which events are not Relevant Events; • what period of time is due against each Relevant Event; • if the A/CA has assessed that the Relevant Event has not caused delay, that the Date for Completion will not be Adjusted. The most appropriate way of reporting this may be as a tabulated analysis given that, once granted, a revision to the Completion Date cannot be taken back. Further reductions or additions to the Completion Date would be dependent upon reductions or expansions in the Works. This changes the process for adjusting completion and elevates the procedural standard to an exacting level akin to that required in the instance of a withholding notice. However, the view has been expressed that Architects presently discharging their duties under the 1998 form of contract ought already to be considering their decisions in such a level of detail – but also that they ought to be stating that consideration in the manner set out in the 2005 version of the contract.	

2.28.4	25.3.2	Clause redrafted: the clause has been significantly reworded and accommodates the incorporation of the Sectional Completion provisions, but does not give rise to new obligations.	
2.28.5	25.3.3	Clause redrafted: the clause has been significantly reworded and accommodates the incorporation of the Sectional Completion provisions, but does not give rise to new obligations.	
2.28.6	25.3.4	**Significant amendment:** the clause has been significantly reworded and gives additional clarity, but does not give rise to new obligations.	
2.29 Relevant Events	25.4	**Major change:** the amendment changes a fundamental process or procedure of the contract compared to JCT 1998.	The list of Relevant Events for SBC/Q 2005 is shorter than that of JCT 1998. Most of the changes are a result of more economic drafting, but CDP provisions are accommodated and there are differences of substance. Certain clauses have been added or made explicit. The Relevant Event concerning impediment, prevention or default now expressly includes action by the Quantity Surveyor or A/CA in clause 2.29.6 (note: this amendment has made clauses 25.4.6.1–2 unnecessary and these have been deleted). If the Planning Supervisor is one of the Employer's Persons, then this will also fall under this clause and thus clause 25.4.17 of JCT 1998 is no longer required. Certain of JCT 1998 Relevant Events have been deleted. Some deletions reflect the removal of particular concepts from SBC/Q 2005, such as the deletion of nomination and the deletion of Performance Specified

SBC/Q 2005 clause and reference title	JCT 1998 clause and reference title	Summary of change	Consequence of change and comments	New action required
			Works. This has rendered clauses 25.4.7 and 25.4.15 of JCT 1998 obsolete. Other changes reflect a shift in risk and a change in the economic climate– for example, the deletion of a Relevant Event for 'Reasonably unforeseeable inability to secure labour' (clause 25.4.10.2) or 'Reasonably unforeseeable inability to secure materials' (clause 25.4.10.2) – at this time these items are perhaps more generally perceived to be in the control of the Contractor.	
Practical Completion, Lateness and Liquidated Damages				
2.30 Practical completion and certificates	17.1	Clause redrafted: the clause has been significantly reworded and accommodates the incorporation of the sectional completion/CDP provisions, but does not give rise to new obligations.	The clause makes provision for certification of completion. It also provides for projects that are being carried out in Sections with the requirement for 'Section Completion Certificates'. Once the project (or all of its Sections) are complete, then the clause provides for a 'Practical Completion Certificate'. This is a new name for what was previously described as a 'Certificate of Practical Completion'. A definition of 'practical completion' has not been included in the SBC/Q 2005, unlike other rival standard forms of contract. It may well be advantageous to both the Employer and the Contractor to agree prior to the commencement of the project the	

			essential requirements to allow practical completion to be granted. It would be expected that such a statement of practical completion would go beyond the completion of construction activity and making good the site, but would also include the preparation and handover of as-built drawings or the handover of maintenance manuals, guarantees and collateral warranties.	
2.31 Non-Completion Certificates	24.1	Clause redrafted: the clause has been significantly reworded and accommodates the incorporation of the Sectional Completion provisions, but does not give rise to new obligations.	This is a requirement for a 'Non-Completion Certificate' (a new defined term) in the event that the Contractor fails to achieve the stated Completion Dates. Previous JCT contracts required the Architect/Contract Administrator to issue a certificate recording the Contractor's failure at the relevant moments, but SBC/Q 2005 gives the certificate name.	Any standard format letters concerning non-completion should be re-examined and updated where necessary to reflect the new terminology.
2.32 Payment or allowance of liquidated damages	24.2.1–2.3	Slight amendment: the clause repeats the text of JCT 1998 with slight changes that improve the drafting or syntax, but the effect is unchanged.		
Partial Possession by Employer				
2.33 Contractor's Consent	18.1	Slight amendment: the clause repeats the text of JCT 1998 with slight changes that improve the drafting or syntax, but the effect is unchanged.		
2.34 Practical Completion Date	18.1.1	Slight amendment: the clause repeats the text of JCT 1998 with slight changes that improve the drafting or syntax, but the effect is unchanged.		

55

SBC/Q 2005 clause and reference title	JCT 1998 clause and reference title	Summary of change	Consequence of change and comments	New action required
2.35 Defects etc. – Relevant Part	18.1.2	Slight amendment: the clause repeats the text of JCT 1998 with slight changes that improve the drafting or syntax, but the effect is unchanged.		
2.36 Insurance – Relevant Part	18.1.3	Slight amendment: the clause repeats the text of JCT 1998 with slight changes that improve the drafting or syntax, but the effect is unchanged.		
2.37 Liquidated damages – Relevant Part	18.1.4	Clause redrafted: the clause has been significantly reworded and gives additional clarity, but does not give rise to new obligations.		
Defects				
2.38 Schedules of defects and instructions	17.2	Slight amendment: the clause repeats the text of JCT 1998 but makes use of new defined terms and makes slight changes, but the effect is unchanged.	The clause retains the same basic provisions as its predecessor, but has an improved layout. The term 'Defects Liability Period' has been superseded by the term 'Rectification Period'. In terms of function, the clause is unchanged.	Update any standard correspondence to accommodate new defined terms and terminology.
2.39 Certificate of Making Good	17.4	**Significant change:** the clause repeats most of the text of JCT 1998, but this is amended in a way that significantly affects the operation of the clause.	SBC/Q 2005 introduces the term 'Certificate of Making Good' which supersedes JCT 1998 term 'Certificate of Making Good Defects'. Although there is a new name for this certificate, the effect is the same. The distinction lies in the removal of the provision that stated that the Contractor was not liable for post-practical completion frost damage.	

Contractor's Design Documents				
2.40 As-built Drawings	Contractor's Designed Portion Supplement – clause 5.10	Slight amendment: the clause repeats the text of JCT 1998, but this is amended to accommodate the incorporation of the Sectional Completion/CDP provisions into SBC/Q 2005.	The provision of as-built drawings for the CDP is a precondition to practical completion of the Works. SBC/Q 2005 makes use of a reordered draft of the amendments provided in the 1998 CDP Supplement.	
2.41 Copyright and use	Contractor's Designed Portion Supplement	**Significant amendment:** the clause imports the text from the CDP Supplement that previously had to be specifically adopted by the Parties.	The extent of the Employer's right to the use the design and material intellectual property is stated. It should be noted that the Employer's rights are explicitly stated to be conditional upon the Contractor having been paid.	
Section 3: Control of the Works				
Access and Representatives				
3.1 Access for Architect/ Contract Administrator	11	Slight amendment: the clause repeats the text of JCT 1998 with slight changes that improve the drafting or syntax, but the effect is unchanged.		
3.2 Person-in-charge	10	Slight amendment: the clause repeats the text of JCT 1998 with slight changes that improve the drafting or syntax, but the effect is unchanged.		
3.3 Employer's Representative	1.9	**Significant change:** the clause repeats most of the text of JCT 1998 but this is amended in a way that significantly affects the operation of the clause.	The clause expressly states what was formerly implied, namely that the Employer is at liberty to remove and reappoint his representative. The clause states that notice of this is to be given to the Contractor – other than this, the clause is identical to clause 1.9 of JCT 1998.	If the A/CA's authority is revoked, then the Employer is to give notice of this to the Contractor.

57

SBC/Q 2005 clause and reference title	JCT 1998 clause and reference title	Summary of change	Consequence of change and comments	New action required
			JCT's notes suggest that the Employer's Representative should not be the A/CA or the Quantity Surveyor.	
3.4 Clerk of Works	12	Slight amendment: the clause repeats the text of JCT 1998 with slight changes that improve the drafting or syntax, but the effect is unchanged.		
3.5 Replacement of Architect/Contract Administrator or Quantity Surveyor	Article 3, Article 4	Slight amendment: the clause repeats the text of JCT 1998 with slight changes that improve the drafting or syntax, but the effect is unchanged.	An improved draft which incorporates text describing processes that had previously been found in the Articles.	
3.6 Contractor's Responsibility	1.5	Clause redrafted: the clause has been significantly reworded and gives additional clarity but does not give rise to new obligations.	The obligation is substantially the same as in JCT 1998, however the drafting is improved. In SBC/Q 2005 two shorter and more comprehensible sentences replace one sentence.	
Sub-letting		**Major change:** the amendment changes a fundamental process or procedure of the contract compared to JCT 1998.	The prime change here is the demise of the provisions for Nominated Sub-contracting. As a consequence there is no longer the need to define sub-contractors as 'domestic'. This has allowed a large amount of text to be removed throughout the contract and this aids comprehension generally.	
3.7 Consent to sub-letting	6.3, 19.2.2	Clause redrafted: the clause has been significantly reworded and gives additional clarity, but does not give rise to new obligations.		

3.8 List in Contract Bills	19.3	Slight amendment: the clause repeats the text of JCT 1998 with slight changes that improve the drafting or syntax, but the effect is unchanged.	The terms are very similar to JCT 1998 but gain an added significance due to the removal of the regime for nominating sub-contractors.
3.9 Conditions of sub-letting	19.4	**Major change:** the amendment changes a fundamental process or procedure of the contract compared to JCT 1998.	This clause is a list of terms that the Contractor must apply to any work he sub-lets. These are substantially the same in effect as in JCT 1998, however there is a major change in that clause 3.9.4 accommodates the new process for the provision of collateral warranties. Clause 3.9.1 provides that the termination of the main contract causes the termination of the sub-contract. Note the revision to the terminology: the contract is no longer 'determined' and is instead 'terminated' – there is no change in effect as a result. Clause 3.9.2.4 requires the Contractor to ensure that collateral warranties are provided within 14 days. Contractor to prepare, arrange execution and provide sub-contract collateral warranties to the Employer within 14 days of a request by the Employer or the A/CA.
Architect/Contract Administrator's Instructions			
3.10 Compliance with instructions	4.1.1 and the Contractor's Designed Portion Supplement	Slight amendment: the clause repeats the text of JCT 1998, but this is amended to accommodate the incorporation of the CDP provisions into SBC/Q 2005.	
3.11 Non-compliance with instructions	4.1.2	Slight amendment: the clause repeats the text of JCT 1998 with slight changes that improve the drafting or syntax, but the effect is unchanged.	

SBC/Q 2005 clause and reference title	JCT 1998 clause and reference title	Summary of change	Consequence of change and comments	New action required
3.12 Instructions to be in writing	4.3.1–4.3.2	Slight amendment: the clause repeats the text of JCT 1998 with slight changes that improve the drafting or syntax, but the effect is unchanged.	Clause 3.12.4 provides that the A/CA may issue a written instruction that tidies up instructions that have not been confirmed in writing. SBC/Q 2005 states that the instruction may be given to have retrospective effect, whereas JCT 1998 provided that the Instruction would be 'deemed to have taken effect on the date on which it was issued otherwise than in writing'. Conspicuous by its absence is any clause stating what is to happen if the Contractor complies with a verbal instruction that is not subsequently confirmed in writing.	
3.13 Provisions empowering instructions	4.2	Clause redrafted: the clause has been significantly reworded, but does not give rise to new rights or obligations.		
3.14 Instructions requiring Variations	13.2.4, 13.2.5	Slight amendment: the clause repeats the text of JCT 1998, but this is amended to accommodate the incorporation of the CDP provisions into SBC/Q 2005.	This clause is expressed in similar terms to JCT 1998, but also provides that a Variation to the Contractor's Designed Portion is a modification to the Employer's Requirements.	Users should be aware of the possible repercussions of changing the Contractor's Designed Portion and be clear as to the extent they wish or intend the Employer's Requirements to be changed. There might be some value in agreeing an amendment to the Employer's Requirements before the issue of the change to the Contractor's Proposals.
3.15 Postponement of work	23.2	No change: the clause repeats the text of JCT 1998.		

3.16 Instructions on Provisional Sums	13.3, 13.3.1	Slight amendment: the clause repeats the text of JCT 1998 with slight changes that improve the drafting or syntax, but the effect is unchanged.	
3.17 Inspection – tests	8.3	Slight amendment: the clause repeats the text of JCT 1998 with slight changes that improve the drafting or syntax, but the effect is unchanged.	
3.18 Work not in accordance with the Contract	8.4	Slight amendment: the clause repeats the text of JCT 1998 with slight changes that improve the drafting or syntax, but the effect is unchanged.	The drafting of this clause has been improved and much of the padding removed, but the processes and obligations are the same as under JCT 1998.
3.19 Workmanship not in accordance with the Contract	8.5	Slight amendment: the clause repeats the text of JCT 1998, but is amended to incorporate or accommodate changes introduced in other clauses.	The power to issue instructions to resolve instances of non-compliant or unsafe working practices is stated here. The changes have brought in the reference to the Health and Safety Plan.
3.20 Executed work	8.2.2	**Significant change:** the clause repeats much of the text of JCT 1998, but also provides for new a process or action from the Parties.	The A/CA is required to advise the Contractor if he is not satisfied with work that is required to be to his satisfaction. SBC/Q 2005 expressly requires that the A/CA states the reasons for his decision (this was not an express requirement in JCT 1998). The requirement that this dissatisfaction be expressed within a reasonable time from the execution of the unsatisfactory work remains.	When advising that the executed work has not achieved the required standard, the A/CA is to provide reasons for his dissatisfaction to the Contractor.
3.21 Exclusion of persons from the Works	8.6	Slight amendment: the clause repeats the text of JCT 1998 with slight changes that improve the drafting or syntax, but the effect is unchanged.	The clause does not permit exclusion for frivolous or vexatious reasons and does not extend to permit exclusion on a whim or on discriminatory grounds.

61

SBC/Q 2005 clause and reference title	JCT 1998 clause and reference title	Summary of change	Consequence of change and comments	New action required
Antiquities				
3.22 Effect of find of antiquities	34.1	Slight amendment: the clause repeats the text of JCT 1998 with slight changes that improve the drafting or syntax, but the effect is unchanged.		
3.23 Instructions on antiquities	34.2	Slight amendment: the clause repeats the text of JCT 1998 with slight changes that improve the drafting or syntax, but the effect is unchanged.		
3.24 Loss and expense arising	34.3.3	Slight amendment: the clause repeats the text of JCT 1998 with slight changes that improve the drafting or syntax, but the effect is unchanged.		
CDM Regulations				
3.25 Undertakings to comply	6A.1, 6A.2, 6A.4	Clause redrafted: the clause has been significantly reworded and gives additional clarity, but does not give rise to new obligations.	SBC/Q 2005 places the obligation to comply with the CDM Regulations at the beginning of the clause. The application of the specific provisions in the sub-clauses will depend on the identity of the Principal Contractor (who may not be the Contractor under the building contract).	
3.26 Appointment of successors	1.6, 6A.3	Slight amendment: the clause repeats the text of JCT 1998 with slight changes that improve the drafting or syntax, but the effect is unchanged.		

Section 4: Payment

Contract Sum and Adjustments

4.1 Work included in Contract Sum	14.1	Slight amendment: the text of JCT 1998 is repeated, but new defined terms are used, slight changes are made and the clause is amended to accommodate changes introduced in other clauses.	The changes to JCT 1998 accommodate the potential for a Contractor's Designed Portion.	
4.2 Adjustment only under the Conditions	14.2	Slight amendment: the clause repeats the text of JCT 1998 with slight changes that improve the drafting or syntax, but the effect is unchanged.		
4.3 Items included in adjustments	30.6.2	Slight amendment: the clause repeats the text of JCT 1998 but is amended to incorporate or accommodate changes introduced in other clauses.	This clause retains the three-category regime for changing the Contract Sum of JCT 1998. The categories are: adjustments, deductions and additions. The clause has been updated – provision is made for the CDP and the removal of the old provisions for nominated sub-contractors has made the clause shorter.	
4.4 Taking adjustments into account	3	Slight amendment: the clause repeats the text of JCT 1998 with slight changes that improve the drafting or syntax, but the effect is unchanged.		
4.5 Final Adjustment	30.6.1.1	Slight amendment: the clause repeats the text of JCT 1998 with slight changes that improve the drafting or syntax, but the effect is unchanged.	This clause is substantially unchanged. The removal of the Nominated Sub-contract provisions allows clause 4.5.1.2 to be shorter than its predecessor (30.6.1.2).	

Certificates and Payments

4.6 VAT	15.2, 15.3	Slight amendment: the clause repeats the text of JCT 1998 with slight changes that improve the drafting or syntax, but the effect is unchanged.	A much shorter clause that performs the same functions as its parallel clause in JCT 1998.	

63

SBC/Q 2005 clause and reference title	JCT 1998 clause and reference title	Summary of change	Consequence of change and comments	New action required
4.7 Construction Industry Scheme (CIS)	30A, 31	**Major change:** the amendment changes a fundamental process or procedure of the contract compared to JCT 1998.	The detailed process applying to payment that was set out in clause 31 of JCT 1998 has been deleted in SBC/Q 2005 and the contract provides that the Contractor is to comply with the CIS if the Employer becomes a 'Contractor' under the scope. Note that the CIS is a different scheme to the 'Scheme for Construction Contracts' that regulates the conduct of adjudication.	The Employer and/or his advisors should be aware of the requirements of the CIS to avoid falling foul of the requirements. Advice from the Employer's accountants or finance department may be required to ascertain whether the Employer is deemed to be a 'Contractor'.
4.8 Advance Payment	30.1.1.6	**Major change:** the amendment changes a fundamental process or procedure of the contract compared to JCT 1998.	The clause itself is unchanged, nor is there any requirement for any kind of advance payment, however users should be aware that the default position has been changed so that if there is an advance payment then there is a presumption in favour of an Advance Payment Bond. If an Advance Payment Bond is not required in respect of an advance payment, then the Contract Particulars must be amended. Note that it is useful for the Employer and the Contractor to agree, prior to the advance payment being made, the schedule or method of repaying the advance payment.	
4.9 Issue of Interim Certificate	30.1.1.1 and 30.1.3	Slight amendment: the clause repeats the text of JCT 1998 with slight changes that improve the drafting or syntax, but the effect is unchanged.		
4.10 Amounts Due in Interim Certificate	30.2	Slight amendment: the clause repeats the text of JCT 1998, but is amended to incorporate or accommodate changes introduced in other clauses.	This clause retains JCT 1998 structure, but the statement of the 'Gross Valuation' has been moved to clause 4.16 of SBC/Q 2005.	

			The removal of the regime for nominated sub-contractors means that one sub-clause has been removed.	
4.11 Interim Valuations	30.1.2.1	Slight amendment: the clause repeats the text of JCT 1998 but is amended to incorporate or accommodate changes introduced in other clauses.		
4.12 Application by Contractor	30.1.2.2	Clause redrafted: the clause has been significantly reworded and gives additional clarity, but does not give rise to new obligations.		
4.13 Interim Certificates – payment	30.1.1.1–30.1.1.5	**Major change:** the amendment changes a fundamental process or procedure of the contract compared to JCT 1998.	The clause has been redrafted and, whilst it is still complicated, it is an improvement on JCT 1998. The Guidance Note to SBC/Q 2005 suggests that the clause has not been changed, but it would be fair to say that the redraft has changed the contract by giving it a better chance of being applied correctly in practice. The final date for payment remains 14 days from the date of issue of the Interim Certificate. Clause 4.13.3 states a positive obligation upon the Employer to write to the Contractor and tell him the sums that he can expect. Users should note that the payment notice under clause 4.13.3 is a valid method of the Employer advising the Contractor that it will pay less than the sum stated in the Interim Certificate. In this respect it is a valid withholding notice, but only if it provides the requisite details as to the sums withheld and the reasons for the withholding. If correctly drafted, then such a notice could stand on its own and no further notice would be required.	The Employer or his advisors should ensure that valid withholding notices are issued within the time limits. If no work has been done in the following month and another Valuation is due, it is necessary to reissue a withholding notice.

65

SBC/Q 2005 clause and reference title	JCT 1998 clause and reference title	Summary of change	Consequence of change and comments	New action required
			Clause 4.13.4 contains the provisions for a withholding notice. The notice required must state the sums, together with the grounds for the withholding. The notice described complies with the format in the Housing Grants, Construction and Regeneration Act 1996 and has not changed. There is no precondition to this clause – thus a notice can be issued under clause 4.13.4 even if no notice was served under clause 4.13.3. Clause 4.13.5 provides that the Employer may only pay a sum different to the sum stated in the Interim Certificate if he issues a certificate under clause 4.13.3, in which case he must pay that sum. The Employer may then only pay a sum different to that in his statement under 4.13.3 if he submits a withholding notice under 4.13.4, in which case he may validly pay the sum stated in the certificate under 4.13.4.	
4.14 Contractor's right of suspension	30.1.4	Clause redrafted: the clause has been significantly reworded and gives additional clarity, but does not give rise to new obligations.	In addition to slightly more contemporary wording, the clause also clarifies the provisions relating to the Contractor's right to suspend Works. The change is slight, but it is suggested that the consequence of this is to reinforce the existing position that if the Contractor's reasons for the suspension were not valid, then it might well be in breach of its other duties and this might itself give rise to a right of termination.	

4.15 Final Certificate – issue and payment	30.8.1–30.8.2	Slight amendment: the clause repeats the text of JCT 1998, but although it makes use of new defined terms and makes slight changes, the effect is unchanged.	The previous mechanisms are restated in slightly sharper terms, but the risk allocation associated with the issue of the Final Certificate is unchanged. In clause 4.15.2 the requirement to issue the Final Certificate remains tied to essentially the same matters as JCT 1998. However, these are updated to take account of some of the new terms (e.g. 'Rectification Period' replaces 'Defects Period').	
Gross Valuation				
4.16 Ascertainment	30.2, 30.2.1	Slight amendment: the clause repeats the text of JCT 1998 with slight changes that improve the drafting or syntax, but the effect is unchanged.		
4.17 Off-site materials and goods	30.3	Clause redrafted: the clause has been significantly reworded and gives additional clarity but does not give rise to new obligations.		
Retention				
4.18 Rules on treatment of Retention	30.5	Clause redrafted: the clause has been significantly reworded and gives additional clarity, but does not give rise to new obligations.	The clause is better expressed, subdivided and is more concise than the previous version in JCT 1998. Local Authority employers will find that the term relating to placing retention money in a separate bank account does not apply to them (clause 4.18.3).	
4.19 Retention Bond	30.4A.1	**Significant change:** the clause repeats much of the text of JCT 1998, but also provides for a new process or action from the Parties.	The clause provides for the option of a Retention Bond as an alternative to deducting cash from the Contractor's payments. The clause now requires that the surety providing the Retention Bond must be satisfactory to the Employer (this is a new requirement and was not stated in JCT 1998).	

67

SBC/Q 2005 clause and reference title	JCT 1998 clause and reference title	Summary of change	Consequence of change and comments	New action required
			Clause 4.19.2 defines the terms 'Retention Bond' and 'Surety' and obliges the Contractor to obtain the Retention Bond from a Surety the Employer deems acceptable. The retention clause is not expressed as an exclusive remedy. Thus if the sum in damages, for which the Retention Bond has been called, exceeds the value of the Retention Bond, then the Employer would be entitled to seek any sums in excess of the value of the Retention Bond from the Contractor by other means including under any performance bond.	
4.20 Retention – rules for ascertainment	30.4.1	**Major change:** the amendment changes a fundamental process or procedure of the contract compared to JCT 1998.	The Retention Percentage is 3% – JCT 1998 provided for retention at a rate of 5%. This is the most significant change to the retention regime. The clause avoids the repetition and references to nomination of the 1998 form and now incorporates reference to sectional completion.	Default Retention Percentage is 3% – any standard documentation should be updated accordingly.
Fluctuations				
4.21 Choice of fluctuation provisions	38, 39, 40	Slight amendment: the clause repeats the text of JCT 1998 with slight changes that improve the drafting or syntax, but the effect is unchanged.	If a Fluctuations Option is to be selected, then this is to be stated in the Contract Particulars.	
4.22 Non-applicability to Schedule 2 Quotation	37.3	Slight amendment: the clause repeats the text of JCT 1998 with slight changes that improve the drafting or syntax, but the effect is unchanged.		

Loss and Expense

4.23 Matters materially affecting regular progress	26.1, 26.2	Clause redrafted: the clause has been significantly reworded and gives additional clarity, but does not give rise to new obligations.	Although abbreviated, the clause is substantially the same in its effect as that in JCT 1998. The loss and expense regime only applies if, and to the extent that, the other terms of the contract do not reimburse the Contractor. In addition, the loss and expense must be due to one of the Relevant Matters or a deferment of possession of the site. Note that the provision remains a loss and/or expense, thus entitlement to one does not necessarily give rise to entitlement to the other.	
4.24 Relevant Matters				
4.24.1	26.2	Slight amendment: the clause repeats the text of JCT 1998 with slight changes that improve the drafting or syntax, but the effect is unchanged.	Note the minor change in terminology from 'matters' to the defined term 'Relevant Matters'.	
4.24.2–4.24.5	26.2.1.1–7	Clause redrafted: the clause has been significantly reworded and gives additional clarity, but does not give rise to new obligations.	SBC/Q 2005 has a shorter list of Relevant Matters than JCT 1998. Clause 4.24.5 addresses many of the points that the previous forms stated as separate subclauses. The 'impediment, prevention or default' referred to in SBC/Q 2005 clause 4.24.2.5 would include restricted site access, non-provision of materials or non-compliance with regulations that were 'matters' that JCT 1998 stated in separate clauses.	
4.25 Amounts ascertained – addition to Contract Sum	26.5	No change: the clause repeats the text of the JCT.		

69

SBC/Q 2005 clause and reference title	JCT 1998 clause and reference title	Summary of change	Consequence of change and comments	New action required
4.26 Reservation of Contractor's rights and remedies	26.6	Slight amendment: the clause repeats the text of JCT 1998 with slight changes that improve the drafting or syntax, but the effect is unchanged.		
Section 5: Variations				
General				
5.1 Definition of Variations	13.1	Slight amendment: the clause repeats the text of JCT 1998, but this is amended to accommodate the incorporation of the CDP provisions into SBC/Q 2005.	The scope of the clause has not been changed but the clause now includes a reference to the Employer's Requirements to facilitate the CDP and deletes the reference to nominated sub-contractors as they no longer exist under SBC/Q 2005.	
5.2 Valuation of Variations and Provisional Sum work	13.4.1.1	Slight amendment: the clause repeats the text of JCT 1998 with slight changes that improve the drafting or syntax, but the effect is unchanged.		
5.3 Schedule 2 Quotation	13A Quotation	Slight amendment: the clause repeats the text of JCT 1998, but is amended to incorporate changes introduced in other clauses.	The text of Clause 13A of JCT 1998 is now split between clause 5.3 and Schedule 2. The process itself is unchanged.	Although not expressly stated, it is reasonable for the Parties to discuss time implications prior to agreement.
5.4 Contractor's right to be present at measurement	13.6	Slight amendment: the clause repeats the text of JCT 1998 with slight changes, but the effect is unchanged.		
5.5 Giving effect to Valuations, Agreements etc.	13.7	Slight amendment: the clause repeats the text of JCT 1998 with slight changes that improve the drafting or syntax, but the effect is unchanged.		

The Valuation Rules		**New definition:** this clause provides a new definition of a term or of a process.	The term 'Valuation Rules' is defined here.	
5.6 Measurable Work	13.5.1–3	Slight amendment: the clause repeats the text of JCT 1998, but this is amended to accommodate the incorporation of CDP provisions into SBC/Q 2005.		
5.7 Daywork	13.5.4	Slight amendment: the clause repeats the text of JCT 1998 with slight changes that improve the drafting or syntax, but the effect is unchanged.		
5.8 Contractor's Designed Portion – Valuation	Contractor's Designed Portion Supplement	Slight amendment: the clause repeats the text of the 1998 CDP Supplement with slight changes that improve the drafting or syntax, but the effect is unchanged.	The clause follows the CDP Supplement very closely and does not deviate from it. Consequently, the effect of the clause is unchanged compared to a JCT 1998 that made use of this supplement. Note the Supplement's prohibition against including loss and expense which is now covered in clause 5.10.2 of SBC/Q 2005.	
5.9 Change of conditions for other work	13.5.5	**Significant change:** the clause repeats most of the text of JCT 1998, but this is amended in a way that significantly affects the operation of the clause.	The clause is largely similar to JCT 1998, but the causal link between the event and the change has been reinforced by the statement that the change is to be a result of the compliance with the instruction or obligation. Although this is a change, it is perhaps a clarification rather than a shift in the balance of risk.	Contractors should expect to prove or demonstrate the direct link between compliance with an instruction and a request for additional monies due to a change of conditions.
5.10 Additional provisions	13.5.7	Slight amendment: the clause repeats the text of JCT 1998 but makes use of new defined terms and makes slight changes but the effect is unchanged.	Clause 5.10.2 makes use of the newly defined term 'Valuation Rules' and also underlines the point that an effect on part of the Works will not give rise to a loss and expense claim under the clause if it is dealt with under another clause.	

SBC/Q 2005 clause and reference title	JCT 1998 clause and reference title	Summary of change	Consequence of change and comments	New action required
Section 6: Injury, Damage and Insurance				
Injury to persons and property				
6.1 Liability of Contractor – personal injury or death	20.1	Clause redrafted: the clause has been significantly reworded and gives additional clarity, but does not give rise to new obligations.	The clause has been moderately redrafted – it makes use of the new definition 'Employer's Persons'. More significantly, the SBC/Q 2005 version of this clause removes the statement that the liability indemnified is 'under statute or common law' – in so doing, it removes any restriction that might conceivably have followed from such a statement.	
6.2 Liability of Contractor – Injury or damage to property	20.2	Slight amendment: the clause repeats the text of JCT 1998 but makes use of new defined terms and makes slight changes, but the effect is unchanged.	Save the use of the new term 'Contractor's Persons', the clause is unaltered.	
6.3 Injury or damage to property – Works and Site Materials excluded	20.3	Slight amendment: the clause repeats the text of JCT 1998 with slight changes that improve the drafting or syntax, but the effect is unchanged.		
Insurance against personal injury and property damage				
6.4 Contractor's insurance of his liability	21.1.1	**Significant change:** the clause repeats most of the text of JCT 1998, but this is amended in a way that significantly affects the operation of the clause.	The parallel clause in JCT 1998 made specific reference to persons with contracts of service and apprenticeships, whereas this version refers to employees only.	Contractors should confirm that apprenticed staff are covered and consider whether

			In response to this users may wish to consider whether their apprentices are employees and the status of self-employed staff.	self-employed individuals are covered.
6.5 Contractor's insurance of liability of Employer	21.2.1	Slight amendment: the text of JCT 1998 is repeated but new defined terms are used, slight changes are made and the clause is amended to accommodate changes introduced in other clauses.		
6.6 Excepted Risks	21.3	No change: the clause repeats the text of the JCT 1998.		
Insurance of the Works			It is essential that the Contract Particulars relating to insurance are completed as not all of the Contract Particulars have a default position.	
6.7 Insurance Options	22.1	Slight amendment: the clause repeats the text of JCT 1998 with slight changes that improve the drafting or syntax, but the effect is unchanged.	The selection of the insurance regime for the project is made in the Contract Particulars and the Options are relocated to Schedule 3. The terms of the options were previously stated in the main terms and conditions of JCT 1998.	
6.8 Related Definitions	1.3, 22.2	**Major change:** the amendment changes a fundamental process or procedure of the contract compared to JCT 1998.	The definition of 'terrorism' is now addressed by direct incorporation of the definitions given in the Terrorism Act 2000. The most significant change in the clause is that the mechanism of loss by terrorism has been de-specified – that is to say, the loss is no longer limited to that caused by 'fire or explosion caused by terrorism', thus greatly widening the range of events that ought to be covered by insurance. In the circumstances, therefore, it might be appropriate to point out that action by protesters seeking to influence the	

SBC/Q 2005 clause and reference title	JCT 1998 clause and reference title	Summary of change	Consequence of change and comments	New action required
			government by force would need to be insured against. This might include environmental protest activity, e.g. directed towards preventing construction activity, or possibly even publicity stunts concerning wider social or legal issues if an element of damage to property were involved. Other than this, the definitions are straightforward. There are a small number of changes to the terms but the effect is minimal. Some widen the scope (e.g. the mechanism for the escape of water is no longer defined in the definition of Specified Perils, which is sensible) others reflect an update in the language used (e.g. 'tempest' is no longer a specified peril, presumably on the grounds that 'storm' is now thought adequately to describe the same Specified Peril).	
6.9 Sub-contractors – Specified Perils cover under Joint Names All Risks Policies	22.3.1	Clause redrafted: the clause has been significantly reworded but does not give rise to new obligations.	This clause ensures that sub-contractors are adequately addressed when the insurance arrangements are put in place – irrespective of whether it is the Employer or the Contractor who is charged with providing the insurance. In essence, sub-contractors will either gain the benefit of the insurance cover as a joint named insured or they are granted a waiver in respect of any subrogation rights. The clause has benefited from being slimmed down by the redraft and the removal of the Nominated Sub-contractor references but nevertheless remains a convoluted clause that requires close attention.	What used to be termed 'domestic sub-contractors' now enjoy the same privileges previously provided only to nominated sub-contractors. All sub-contractors should be considered as insurance is put in place for the project including those not yet appointed.

6.10 Terrorism Cover – non-availability – Employer's options	22A.5, 22B.4.1, 22C.5.1	Slight amendment: This clause repeats the text of JCT 1998, but this is amended to incorporate or accommodate changes introduced in other clauses.	The insuring Party has to advise the other Party that Terrorism Cover is about to lapse. At this point the Employer has an option to carry on without the insurance or terminate the contract.	
CDP Professional Indemnity Insurance				
6.11 Obligation to insure	N/A	**Major change:** This is a wholly new clause for the SBC/Q 2005.	An obligation for the Contractor to maintain Professional Indemnity Insurance in respect of the Contractor's Designed Portion and an obligation to provide proof upon request. This is a new requirement for SBC/Q 2005 – neither JCT 1998 nor the 1998 Contractor's Designed Portion Supplement made such provision. One questions whether this will result in a great change in practice given that many Contractors already maintain such insurances.	A Contractor undertaking a CDP is to secure Professional Indemnity Insurance.
6.12 Increased cost and non-availability	N/A	**Major change:** This is a wholly new clause for the SBC/Q 2005.	Although notice is to be given of the fact that insurance is uneconomic, the Contract provides no specific resolution in the event that the insurance becomes uneconomic.	Any notices to the Employer should be provided formally as stipulated elsewhere in the Contract.
Joint Fire Code – Compliance				
6.13 Application of clause	22FC.1	No change: the clause repeats the text of the JCT.		
6.14 Compliance with Joint Fire Code	22FC.2	Slight amendment: the clause repeats the text of JCT 1998 but makes use of new defined terms and makes slight changes, but the effect is unchanged.		

75

SBC/Q 2005 clause and reference title	JCT 1998 clause and reference title	Summary of change	Consequence of change and comments	New action required
6.15 Breach of Joint Fire Code – Remedial measures	22FC.3	Slight amendment: the clause repeats the text of JCT 1998 with slight changes that improve the drafting or syntax, but the effect is unchanged.	In comparison to JCT 1998 the only change of substance is the removal of the defined term 'Remedial Measures Completion Date'. It is replaced by the recognition in the clause that the requirements stated by the insurer will prevail.	
6.16 Joint Fire Code – amendment/revisions	22FC.5	Slight amendment: the clause repeats the text of JCT 1998 with slight changes that improve the drafting or syntax, but the effect is unchanged.		
Section 7: Assignment, Third Party Rights and Collateral Warranties				
Assignment				
7.1 General	19.1.1	Slight amendment: the clause repeats the text of JCT 1998 but makes use of new defined terms and makes slight changes, but the effect is unchanged.	Like JCT 1998, SBC/Q 2005 contract provides no free right of assignment – any assignment must have the consent of the other Party. Note also that assignment without consent is also a termination event.	
7.2 Rights of enforcement	19.1.2	Slight amendment: the clause repeats the text of JCT 1998 but makes use of new defined terms and makes slight changes, but the effect is unchanged.		
Clauses 7A to 7F – Preliminary	N/A		Part 2 of the Contract Particulars deals specifically with third party rights and Collateral Warranties – it is essential that this is completed as the default position is that there are to be no such rights unless there is a specific opt-in.	Complete the third party rights section of the Contract Particulars if there are to be any such rights.

			The rights themselves are set out in Schedule 5 of the contract and all of the rights are optional.	
7.3 References	N/A	**Major change:** the amendment changes a fundamental process or procedure of the contract compared to JCT 1998.	This term allows the incorporation of further documents or schedules of rights that may be appended to the contract.	
7.4 Notices	N/A	**Major change:** the amendment changes a fundamental process or procedure of the contract compared to JCT 1998.	The requirements for giving proper notice under this clause are stated here. It is suggested that if a JCT standard form Collateral Warranty were to be amended, then the warranty would not be a 'specified JCT Collateral Warranty' and the amended form would have to accompany the notice.	To provide a notice the Parties should not use the normal post; instead, documents should be sent by recorded or special delivery, or by courier, and a signature taken at point of delivery to prove receipt.
7.5 Execution of Collateral Warranties	N/A	**Major change:** the amendment changes a fundamental process or procedure of the contract compared to JCT 1998.	This clause ensures that the Collateral Warranties required follow the main contract so that a Warranty by deed is not required where the main contract is under hand, and vice versa. Most readers will be aware that the limitation period for an action for a breach of a contract executed as a simple contract is six years and that the period is 12 years for a deed.	
Third Party Rights from Contractor				
7A Rights for Purchasers and Tenants	N/A	**Major change:** This is a wholly new clause for the SBC/Q 2005 and the balance of risk is changed as a result.	The Contract Particulars must state this requirement. The rights vest at the time the Employer gives the notice to the Contractor. The notice must comply with 7.4 by being in writing and given by actual delivery (i.e. delivered by hand rather than mailed), or by special or recorded delivery. Ordinary post or fax does not suffice. The notice	If Purchaser and Tenant rights are required, then this requirement is to be stated in the Contract Particulars before the contract is executed.

SBC/Q 2005 clause and reference title	JCT 1998 clause and reference title	Summary of change	Consequence of change and comments	New action required
			must also comply with 7A.1, which requires the identity of the person acquiring the rights and their interest to be stated – that is to say, the nature of their proximity or relevance to the Works. Clause 7A.2 notes that the Parties are free to continue to exercise their own rights (such as termination, waiver or settlement) under the contract even though Purchasers and Tenants (P&T) may have rights too. The Parties can agree to amend the rights that will be given, however clause 7A.3 does prevent the Parties from changing the rights of the Purchasers and Tenants once the P&T Rights have been vested.	
7B Rights for a Funder	N/A	**Major change:** This is a wholly new clause for the SBC/Q 2005 and the balance of risk is changed as a result.	The Particulars must state this requirement. The rights vest at the time the Employer gives the notice to the Contractor. The notice must comply with 7.4 by being in writing and given by actual delivery (i.e. delivered by hand rather than mailed), or by special or recorded delivery. Ordinary post or fax does not suffice. The notice must also comply with 7B.1 which requires the identity of the Funder acquiring the rights. The clause permits the Parties to operate the contract without needing to get the consent of the Funder – thus Variation and settlement of items in dispute and waiver of terms are not restricted. However unlike the P&T Rights, the rights of the Funder bind both the Parties in some respects. Specifically, clause 7B.2 prevents either Party from rescinding the contract.	If Funder's rights are required, then this requirement is to be stated in the Contract Particulars before the contract is executed.

Collateral Warranties				
7C Contractor's Warranties – Purchasers and Tenants	N/A	**Major change:** This is a wholly new clause for the SBC/Q 2005 and the balance of risk is changed as a result.	The Particulars must state that Collateral Warranties for Purchasers and Tenants are a requirement. The obligation to procure the Warranties arises at the time the Employer gives the notice to the Contractor. The notice must comply with clause 7.4 by being in writing and given by actual delivery (i.e. delivered by hand rather than mailed), or by special or recorded delivery. Ordinary post or fax does not suffice. The notice must also enclose the terms of the Warranty if it is different from or has amendments to the JCT Collateral Warranty (7.4). The Contractor must enter into the Warranty within 14 days of receiving the notice.	If Contractor's Collateral Warranties are required in favour of Purchasers and Tenants rights, then this is to be stated in the Contract Particulars before the contract is executed.
7D Contractor's Warranty – Funder	N/A	**Major change:** This is a wholly new clause for the SBC/Q 2005 and the balance of risk is changed as a result.	The Particulars must state that Collateral Warranties for a Funder are a requirement. The obligation to procure the Warranties arises at the time the Employer gives the notice to the Contractor. The notice must comply with 7.4 by being in writing and given by actual delivery (i.e. delivered by hand rather than mailed), or by special or recorded delivery. Ordinary post or fax does not suffice. The notice must also enclose the terms of the Warranty if it is different from or has amendments to the JCT Collateral Warranty (7.4). The Contractor must enter into the Warranty within 14 days of receiving the notice.	If Contractor's Collateral Warranties are required in favour of the Funder, then this requirement is to be stated in the Contract Particulars before the contract is executed.

SBC/Q 2005 clause and reference title	JCT 1998 clause and reference title	Summary of change	Consequence of change and comments	New action required
7E Sub-contractor's Warranties – Purchasers and Tenants/Funder	N/A	**Major change:** This is a wholly new clause for the SBC/Q 2005 and the balance of risk is changed as a result.	The Particulars must state that Collateral Warranties are required from Sub-contractors in favour of Purchasers and Tenants. The obligation to provide the Collateral Warranty takes effect upon the receipt of a notice that complies with clause 7.4 by being in writing and given by actual delivery (i.e. delivered by hand rather than mailed), or by special or recorded delivery. Ordinary post or fax does not suffice. The notice also has to state which sub-contractor the notice is required from, in whose favour and in which form. The notice must also enclose the terms of the Warranty if it is different from or has amendments to the JCT Collateral Warranty (7.4). The Contractor must achieve the execution of the Warranty by its Sub-Contractor within 21 days of the Employer's notice.	If Sub-contractor's Collateral Warranties are required in favour of the Purchasers and Tenants then this requirement is to be stated in the Contract Particulars before the contract is executed.
7F Sub-contractor's Warranties – Employer	N/A	**Major change:** this is a wholly new clause for the SBC/Q 2005 and the balance of risk is changed as a result.	The Particulars must state that Collateral Warranties are required from Sub-contractors in favour of the Employer. The obligation to provide the Collateral Warranty takes effect upon the receipt of a notice that complies with clause 7.4 by being in writing and given by actual delivery (i.e. delivered by hand rather than mailed), or by special or recorded delivery. Ordinary post or fax does not suffice. The notice also has to state which Sub-contractor the notice is required from, in whose favour and in which form. The	If the Employer requires Sub-contractor's Collateral Warranties in his favour then this requirement is to be stated in the Contract Particulars before the contract is executed.

notice must also enclose the terms of the Warranty if it is different or has amendments to the JCT Collateral Warranty (7.4).

The Contractor must achieve the execution of the Warranty by its Sub-contractor within 21 days of the Employer's notice. The clause specifically considers that the Warranty may be provided in different terms. The Employer's approval of the amendments is required, however he cannot unreasonably delay or withhold consent.

The result is that small-scale changes would ultimately have to be accepted by the Employer. If the Employer states specific requirements in the Contract Particulars, then it is suggested that it would not be unreasonable to withhold consent if the warranty proposed does not satisfy the requirements.

Section 8: Termination

Generally:
The first change noted will be that a contract is 'terminated' rather than 'determined'. In general the clause is more concise and a good deal of repetition has been avoided. In operation and effect the contract is very similar to JCT 1998. Many of the changes relate to insolvency and the JCT's own guidance note is particularly helpful on the practical aspects of the new clause.

Most significantly, a notice will always be required to terminate the contract, even in the event of insolvency, as there is no longer automatic termination in the event of insolvency.

It should be noted that the provision for the payment of Sub-contractors directly has been omitted, as has the arrangement for novation or continuation.

As with JCT 1998, both the Employer and the Contractor should be extremely careful to correctly terminate the contract, should that be necessary.

The termination procedure is exact and technical, and if either Party terminates incorrectly and does not perform their obligations under the contract, then that Party might find that they have repudiated the contract.

Note well that a notice must now be submitted to terminate the contract in even the event of insolvency.

SBC/Q 2005 clause and reference title	JCT 1998 clause and reference title	Summary of change	Consequence of change and comments	New action required
General				
8.1 Meaning of Insolvency	27.3.1	**Major change:** the amendment changes a fundamental process or procedure of the contract compared to JCT 1998.	The JCT Guidance Note explains that the new clause has been worked up with the assistance of the Association of Insolvency Practitioners. The clause stating the meaning of insolvency for the purpose of this contract contains many of the elements of the previous version of the contract, however it appears to be a wider definition. Clause 8.1.1 in particular is of broad reach and would catch many of the preparatory stages of insolvency.	
8.2 Notices under section 8	27.1, 27.2.4	Clause redrafted: the clause has been significantly reworded and gives additional clarity but does not give rise to new obligations.	The mechanism for giving notice is the same as the previous version of the contract and the same as that stated under the warranty clauses. Notices of termination cannot be given merely to annoy or vex, nor can they be given unreasonably. The termination notice takes effect upon the receipt of a notice that complies with clause 8.2 by being in writing and given by actual delivery (i.e. delivered by hand rather than mailed), or by special or recorded delivery. Ordinary post or fax does not suffice. The assumption of delivery on the second Business Day can be rebutted if a Party can demonstrate that the notice was given sooner, or indeed that it was not given at all.	
8.3 Other rights, reinstatement	27.8, 28.5	Slight amendment: the clause repeats the text of JCT 1998 with slight changes that improve the drafting or syntax, but the effect is unchanged.	The clause expressly states that the Contractor may be reinstated if both Parties agree. If the circumstances are favourable, there is no need to prepare a completely new contract.	

Termination by Employer

8.4 Default by Contractor	27.2.1–5	Slight amendment: the clause repeats the text of JCT 1998 with slight changes that improve the drafting or syntax, but the effect is unchanged.		
8.5 Insolvency of Contractor	27.3.2	**Major change:** the rights of the Parties have been changed by this amendment compared to JCT 1998.	If the Contractor is Insolvent, the Employer can terminate immediately by notice. The onus is therefore upon the Employer to act swiftly and decisively in the instance of the insolvency of the Contractor. This is a major change compared to JCT 1998, where determination occurred immediately and without any action from the Employer upon the Contractor's insolvency in almost all cases. The Contractor is obliged to tell the Employer if he is approaching insolvency. Notwithstanding this and other legal obligations that apply to companies approaching insolvency, the workability of this clause has be questioned.	Notice must be given to the Contractor by the Employer if he wishes to terminate the contract in the event of the Contractor's insolvency.
8.6 Corruption	27.4	Clause redrafted: the clause has been significantly reworded, but does not give rise to new obligations.	A concise statement that the Contractor's corruption on any project will permit the Employer to terminate the contract. The clause also accommodates the specific needs of local authorities.	
8.7 Consequences of termination under clauses 8.4 to 8.6	27.6.1, 27.6.3	Clause redrafted: the clause has been significantly reworded and gives additional clarity, but does not give rise to new obligations.	The Employer may have another Contractor carry on and complete the Works and may have supply contracts assigned to him.	
8.8 Employer's decision not to complete the Works	27.7.1	Slight amendment: the clause repeats the text of JCT 1998 but makes use of new defined terms and makes slight changes, but the effect is unchanged.	There is a slight change found in clause 8.8.2 to state that the Employer can only deduct sums for which the Contractor is liable (this is a change in the interests of clarity and does not extend the ambit of the clause).	

83

JCT 2005 Standard Building Contract With Quantities (SBC/Q 2005)

SBC/Q 2005 clause and reference title	JCT 1998 clause and reference title	Summary of change	Consequence of change and comments	New action required
Termination by Contractor				
8.9 Default by the Employer	28.2.1.1–4	Slight amendment: the clause repeats the text of JCT 1998 with slight changes that improve the drafting or syntax, but the effect is unchanged.		
8.10 Insolvency of Employer	28.3.1	Clause redrafted: the clause has been significantly reworded, but does not give rise to new obligations.		
8.11 Termination by either Party	28A	Clause redrafted: the clause has been significantly reworded and gives additional clarity, but does not give rise to new obligations.		
8.12 Consequences of termination under clauses 8.9 to 8.11	28A.2–6	**Major change:** the amendment changes a fundamental process or procedure of the contract compared to JCT 1998.	After termination, an account stating the sums due is to be compiled. The distinction between JCT 1998 and SBC/Q 2005 is that clause 28A.2 of JCT 1998 required the Employer to prepare the account, whereas under SBC/Q 2005 the Contractor prepares the account if the contract was terminated under clause 8.9 (termination for default by the Employer) or clause 8.10 (Employer's insolvency). Note that the Contractor is obliged to prepare this account as soon as reasonably practicable and in any case not more than two months from termination.	If terminated under 8.9 or 8.10, the Contractor is to draw up the account.

Section 9: Settlement of Disputes				
9.1 Mediation	N/A	**Major change:** this is a wholly new clause for the SBC/Q 2005.	The clause is a statement that the Parties may have their differences resolved by mediation – it is not an obligation that they must do so, nor that they must even consider doing so. This statement is a new term for SBC/Q 2005 and also reflects a useful process that has been supported by the courts. Mediation may, in the particular circumstances of a dispute, be a sensible and cost-effective way to resolve a dispute. Certainly the courts regularly endorse mediation and are diverting many cases in this direction and penalising in costs those who do not consider it.	Give serious consideration to mediating any dispute, particularly if the project is in its early stages and there are sound reasons for avoiding adversarial processes.
9.2 Adjudication	41A	**Major change:** this is a wholly new clause for the SBC/Q 2005 and the balance of risk is changed as a result.	The adjudication procedure set out at length in JCT 1998 has been deleted. The SBC/Q 2005 adopts the Scheme set out in The Scheme for Construction Contracts (England and Wales) Regulations 1998 ('the Scheme'). The operation of the Scheme is slightly amended. Firstly, by the provision that the Parties may identify their own Adjudicator or Adjudicator Nominating Body if the necessary amendment is made in the Contract Particulars (this is the same as the Adjudication rules under the JCT 1998 contract). Secondly, clause 9.2.2 also departs from the Scheme and provides that an Adjudicator of a dispute concerning opening up and testing under clause 3.18.4 must either have specialist knowledge or must appoint a specialist advisor. Users should familiarise themselves with the Scheme for Construction Contracts.	The Scheme for Construction Contracts should be (re)read before contemplating any adjudication. Terms of the Adjudicator's appointment must be agreed with the Adjudicator.

85

SBC/Q 2005 clause and reference title	JCT 1998 clause and reference title	Summary of change	Consequence of change and comments	New action required
			It is not the purpose of the present exercise to provide an analysis of the Scheme, its processes nor its relative merits and demerits compared to JCT 1998 procedure. In addition to the removal of the JCT's adjudication process, it is no longer a condition that the Adjudicator has to accept the JCT Adjudication Agreement that provided terms between the Parties and the Adjudicator – all references to it have been removed. This is a positive change that has removed a potential difficulty in the Adjudicator selection process. Problems occasionally arose if Adjudicators were not willing to use the JCT Adjudication Agreement.	
Arbitration				
9.3 Conduct of arbitration	41B.6	**Major change:** this is a wholly new clause for SBC/Q 2005.	Resolution of disputes by arbitration is not the default position under the SBC/Q 2005: unless the Parties make the necessary changes in the Contract Particulars these clauses will not apply. The Parties to a contract that does not adopt arbitration may still have a dispute arbitrated if they both agree. In such a circumstance the Parties would not be arbitrating 'pursuant to Article 8', thus unless they also agreed to accept them, these clauses would not apply and the Parties would also have to agree the appropriate arbitration rules.	A decision is required prior to the formation of the contract as to whether arbitration or litigation is required. The relative merits of each method should be considered.

9.4 Notice of reference to arbitration	41B.1.1	Clause redrafted: the clause has been significantly reworded but does not give rise to new obligations.	The method and process for giving notice of arbitration is stated here. The distinction is that JCT 1998 reproduced the relevant part of Rule 2.1 of the Construction Industry Model Arbitration Rules (CIMAR) – SBC/Q 2005 does not.	
9.5 Powers of Arbitrator	41B.2	Slight amendment: the clause repeats the text of JCT 1998 with slight changes that improve the drafting or syntax, but the effect is unchanged.		
9.6 Effect of Award	41B.3	Slight amendment: the clause repeats the text of JCT 1998 with slight changes that improve the drafting or syntax, but the effect is unchanged.		
9.7 Appeal – questions of law	41B.4	Slight amendment: the clause repeats the text of JCT 1998 with slight changes that improve the drafting or syntax, but the effect is unchanged.		
9.8 Arbitration Act 1996	41B.5	Slight amendment: the clause repeats the text of JCT 1998 with slight changes that improve the drafting or syntax, but the effect is unchanged.		
SCHEDULES				
Schedule 1: Contractor's Design Submission Procedure	Contractor's Designed Portion Supplement	**Major change:** this is a wholly new clause for the SBC/Q 2005.	This design submission procedure applies only to the CDP. The design submission and approval procedure was not present in either JCT 1998 or the 1998 CDP Supplement. This Schedule will be familiar to those who have used the JCT Major Projects Form of 2003. In keeping with the latest draft of SBC/Q 2005, the new design submission procedure is fairly clearly drafted.	The process only applies if there is a CDP. If there is a CDP, the Contractor is obliged to comply with the procedure irrespective of the action or inaction of the A/CA.

87

SBC/Q 2005 clause and reference title	JCT 1998 clause and reference title	Summary of change	Consequence of change and comments	New action required
			However, the contract does not state a defined period of days as the time for the Contractor to provide its 'Contractor's Design Document'. Instead this is to be submitted 'in sufficient time to allow any comments … to be incorporated prior to … being used'. This element of conditionality might lead to difficulty as Parties may have a different idea as to what 'sufficient time' may be. A practical solution to this would be for the Employer to state a time-period for the provision of the Documents in his Requirements. Once the Contractor's Design Documents have been received by the A/CA then there will be a defined outcome in a defined period. If the A/CA fails to act in time, then the documents are taken to be approved. Users should note clause 8.3, which provides that the Contractor's design obligation is not diluted by any involvement or observation by the A/CA in connection with the Contractor's Design Submission Procedure. That is to say that the Contractor must comply with the procedure even if the A/CA elects not to contribute to the process. The procedure ensures that the Contractor's obligations are clear in either instance.	
Schedule 2: Schedule 2 Quotation	13A Variation Instruction – Contractor's Quotation in compliance with the instruction			

Submission of Quotation	13A1.1	Slight amendment: the clause repeats the text of JCT 1998 with slight changes that improve the drafting or syntax, but the effect is unchanged.	The process of the submission is the same as under JCT 1998. Such changes as are made improve the drafting and remove the provisions relating to nominated sub-contractors.	
Content of the Quotation	13A.2	Slight amendment: the clause repeats the text of JCT 1998 with slight changes that improve the drafting or syntax, but the effect is unchanged.		
Acceptance of the Quotation	13A3.1	Slight amendment: the clause repeats the text of JCT 1998 with slight changes that improve the drafting or syntax, but the effect is unchanged.		
Quotation not accepted	13A.4	Slight amendment: the clause repeats the text of JCT 1998 with slight changes that improve the drafting or syntax, but the effect is unchanged.		
Costs of Quotation	13A.5	No change: the clause repeats the text of JCT 1998.		
Restriction on use of Quotation	13A.6	Slight amendment: the clause repeats the text of JCT 1998 with slight changes that improve the drafting or syntax, but the effect is unchanged.		
Time Periods	13A.7	Slight amendment: the clause repeats the text of JCT 1998 but is amended to incorporate or accommodate changes introduced in other clauses.		
Schedule 3: Insurance Options				
Insurance Option A				
(New Buildings – All Risks Insurance of the Works by the Contractor)				

89

SBC/Q 2005 clause and reference title	JCT 1998 clause and reference title	Summary of change	Consequence of change and comments	New action required
A.1 Contractor to take out and maintain Joint Names Policy	22A.1	Slight amendment: the clause repeats the text of JCT 1998 with slight changes that improve the drafting or syntax, but the effect is unchanged.		
A.2 Insurance Documents – failure by Contractor to Insure	22A.2	**Significant change:** the clause repeats most of the text of JCT 1998, but this is amended in a way that significantly affects the operation of the clause.	If paragraph A.1 and A.2 are being applied (as opposed to paragraph A.3) then insurance documents must now be lodged with the Employer and not the Contractor – otherwise the process is unchanged.	The Contractor is to send the insurance policy, periodic insurance payment receipts and policy endorsements to the Employer.
A.3 Use of Contractor's Annual Policy – as alternative	22A.3.1	Clause redrafted: the clause has been significantly reworded, but does not give rise to new obligations.	Paragraph A.1 and A.2 will apply if the Contractor is not able to meet its obligation from its 'normal' insurance policies (i.e. those that it might possess in the general course of its business as a Contractor). Clause 22A.3.2 of JCT 1998 stated that the Contractor had to arrange specific insurances if its general policy lapsed – this point is now subsumed within clause A.3 of SBC/Q 2005 and the separate clause is no longer necessary.	
A.4 Loss or damage, insurance claims and Contractor's obligations	22A.4.1–5	Slight amendment: the clause repeats the text of JCT 1998 with slight changes that improve the drafting or syntax, but the effect is unchanged.		
A.5 Terrorism Cover – premium rate changes	22A.5.4.1	Slight amendment: the clause repeats the text of JCT 1998 with slight changes that improve the drafting or syntax, but the effect is unchanged.		

Insurance Option B				
(New Buildings – All Risks Insurance of the Works by the Employer)	22B.1	Slight amendment: the clause repeats the text of JCT 1998 with slight changes that improve the drafting or syntax, but the effect is unchanged.		
B.1 Employer to take out and maintain a Joint Names Policy	22B.2	Slight amendment: the clause repeats the text of JCT 1998, but this is amended to accommodate the incorporation of the sectional completion provisions into SBC/Q 2005.		
B.2 Evidence of Insurance	22B.2	**Significant change:** the clause repeats much of the text of JCT 1998 but also provides for new process or action.	The paragraph makes provision for both private and local authority Employers. A local authority is obliged to provide a cover certificate rather than documentary evidence of the insurance.	
B.3 Loss or damage, insurance claims, Contractor's obligations and payment to the Employer	22B.3	Slight amendment: the clause repeats the text of JCT 1998 but is amended to incorporate or accommodate changes introduced in other clauses.	Paragraph B.3.2 refers to clause 6.10.4.2 and thus accommodates the fact that different rules are to apply to recoveries following loss or damage if Terrorism Cover is not available and the Employer has instructed the Contractor to continue with the Works despite this.	
Insurance Option C				
Insurance by the Employer of Existing Structures and Works in or Extensions to them				
C.1 Existing structures and contents – Joint Names Policy for Specified Perils	22C.1	Clause redrafted: the clause has been significantly reworded and gives additional clarity, but does not give rise to new obligations.	The VAT provisions located at the end of the clause have been relocated.	
C.2 The Works – Joint Names Policy for All Risks	22C.2	Clause redrafted: the clause has been significantly reworded and gives additional clarity, but does not give rise to new obligations.	The VAT provisions located at the end of the clause have been relocated.	

91

JCT 2005 Standard Building Contract With Quantities (SBC/Q 2005)

SBC/Q 2005 clause and reference title	JCT 1998 clause and reference title	Summary of change	Consequence of change and comments	New action required
C.3 Evidence of Insurance	22C.3	Slight amendment: the clause repeats the text of JCT 1998 but is amended to incorporate or accommodate changes introduced in other clauses.	The paragraph makes provision for both private and local authority employers. A local authority is obliged to provide a cover certificate rather than documentary evidence of the insurance.	
C.4 Loss or damage to Works – insurance claims and Contractor's obligations	22C.4	Slight amendment: the clause repeats the text of JCT 1998, but is amended to incorporate or accommodate changes introduced in other clauses.	Paragraph C.4.2 refers to clause 6.10.4.2 and thus accommodates the fact that different rules are to apply to recoveries following loss or damage if Terrorism Cover is not available and the Employer has instructed the Contractor to continue with the Works despite this.	
Schedule 4: Code of Practice	Code of Practice: Referred to in Clause 8.4.4	No change: the clause repeats the text of the JCT.	Save very small changes (one or two words have been substituted) the Code of Practice has not been changed in any respect and its effect is unchanged.	
Schedule 5: Third Party Rights				
Part 1: Third Party Rights for Purchasers and Tenants			*Generally:* SBC/Q is the first JCT contract comprehensively to address the rights that can accrue to Purchasers, Tenants and Funders without the need to include supplements or other pro forma warranties. Users must be aware that the third party rights regime of SBC/Q 2005 is optional, and consequently the Parties must state which provisions they adopt in Part 2 of the Contract Particulars. A fair amount of thought is necessary to select options that are appropriate to the particular project being considered.	

			Comparison is made to the 2001 Standard Form Agreement for Collateral Warranty for Purchasers and Tenants (Ref. MCWa/P&T) that is replaced by Schedule 5 of SBC/Q 2005. References in Column 2 are references to clauses in that 2001 Standard Form Agreement.	
1.1	2001 Collateral Warranty Recital C and 1(a)	Slight amendment: the clause repeats the text of the 2001 P&T Collateral Warranty with slight changes that improve the drafting or syntax, but the effect is unchanged.		
1.2	1(b)	Slight amendment: the clause repeats the text of the 2001 P&T Collateral Warranty with slight changes that improve the drafting or syntax, but the effect is unchanged.	This paragraph underpins the point that the Contractor is only liable under clause 1.1.2 if the Contract Particulars are completed to state that this is to be the case.	
1.3	1(c)	Slight amendment: the clause repeats the text of the 2001 P&T Collateral Warranty with slight changes that improve the drafting or syntax, but the effect is unchanged.	The paragraph establishes that the Contractor is responsible for only those losses that it is just and equitable to require the Contractor to pay. The assumptions of the elements of the liability that are (theoretically) met by the other Parties are key and are stated in the sub clauses paragraphs 1.3.1–3.	
1.4	1(d)	No change: the clause repeats the 2001 P&T Collateral Warranty.		
1.5	1(e)	No change: the clause repeats the 2001 P&T Collateral Warranty.		
2	2	Slight amendment: the clause repeats the text of the 2001 P&T Collateral Warranty with slight changes that improve the drafting or syntax, but the effect is unchanged.	The 'prohibited materials' themselves are stated in the industry standard document produced by Ove Arup.	

SBC/Q 2005 clause and reference title	JCT 1998 clause and reference title	Summary of change	Consequence of change and comments	New action required
3	3	No change: the clause repeats the 2001 P&T Collateral Warranty.	Although SBC/Q 2005 gives the Purchasers and Tenants rights it does not give them any formal means of directing the Contractor or of giving instructions.	
4	4	**Significant change:** the clause repeats most of the text of the 2001 P&T Collateral Warranty but this is amended in a way that significantly affects the operation of the clause.	The Purchaser or Tenant will only be able to be conveyed rights and licenses to the CDP documents if it is the Purchaser or Tenant of a part of the site that falls within the Contractor's Designed Portion. This is a more limited right than that stated in the 2001 Collateral Warranty which was less proscriptive about who would be able to acquire the rights. Note that as with the 2001 P&T Collateral Warranty Form, P&T Rights under this clause are subject to the Contractor having been fully paid (not just sums due in respect of the CDP).	
5	5	**Significant change:** the clause automatically applies a clause that previously had to be specifically adopted by the Parties.	The clause has been redrafted – this reflects its inclusion into the SBC/Q 2005 main contract as opposed to being a stand-alone document to be used for other forms of JCT contract such as the Intermediate Form or With Contractor's Design Form. The reference in the Contract Particulars states that the default position in the Contract Particulars is that the Professional Indemnity Insurance will be for the stated value in aggregate (rather than for each claim arising out of an event) unless a contrary position is stated.	State the basis on which Professional Indemnity Insurance is required.
6	6	No change: the clause repeats the 2001 P&T Collateral Warranty.		

7	7	Slight amendment: the clause repeats the text of the 2001 P&T Collateral Warranty with slight changes that improve the drafting or syntax, but the effect is unchanged.		
8	8	Clause redrafted: the clause has been significantly reworded, but does not give rise to new obligations.		
9	9	No change: the clause repeats the 2001 P&T Collateral Warranty.		
10	11	Clause redrafted: the clause has been significantly reworded, but does not give rise to new obligations.		
Part 2: Third Party Rights for a Funder			The Contractor's liability to a Funder is stated here – it is potentially far greater than that owed to either a Purchaser or Tenant. The key distinction in the scope of the rights themselves is the provision for the Funder to activate its rights to remove the Employer and take over the administration of the project (often termed activating 'step-in rights'). This is a wider set of obligations than those granted to Purchasers and Tenants as those rights only take effect from practical completion. This reflects the probable involvement of the Funder before and during the construction period as well as the relative strength of a Funder. Comparison is made to 2001 Standard Form Agreement for Collateral Warranty for a Funder (Ref. MCWa/F) that is replaced by Schedule 5 Part 2 of SBC/Q 2005. References in Column 2 are references to clauses in that 2001 Standard Form Agreement.	

95

SBC/Q 2005 clause and reference title	JCT 1998 clause and reference title	Summary of change	Consequence of change and comments	New action required
1	1	Slight amendment: the clause repeats the text of the 2001 P&T Collateral Warranty with slight changes that improve the drafting or syntax, but the effect is unchanged.	The only difference in comparison to the 2001 Collateral Warranty is the accommodation of the mechanism of the rights being provided under the Contracts (Rights of Third Parties) Act 1999 if that method of gaining the rights is preferred.	
2	2	No change: the clause repeats the JCT 2001 Funder Collateral Warranty.		
3	3	No change: the clause repeats the JCT 2001 Funder Collateral Warranty.		
4	4	No change: the clause repeats the JCT 2001 Funder Collateral Warranty.		
5	5	No change: the clause repeats the JCT 2001 Funder Collateral Warranty.	Note that the roles of the Employer's Persons are not changed as a result of such a step-in notice. The A/CA and Quantity Surveyor remain in post and they retain their duties and powers under the contract although ultimately the Funder may be at liberty to make changes.	
6	6	No change: the clause repeats the JCT 2001 Funder Collateral Warranty.		
7	7	No change: the clause repeats the JCT 2001 Funder Collateral Warranty.	This clause states the consequence of the Funder stepping into the Employer's place. Simply stated, the Funder assumes all of the liability that had previously been the Employer's – specifically, all sums due or that will fall due and any obligation to pay sums due at the date of the notice. This clause provides the Contractor with a degree of protection. If the Funder activates its rights it must pay any sums outstanding to the Contractor – of course there may be	

			some debate as to what sums are due, but the existing team of consultants are likely to be on hand to assist.	
8	8	Clause redrafted: the clause has been significantly reworded, but does not give rise to new obligations.	The Funder may use the CDP material in the same way and with the same limitations as the Employer. This is, however, subject to the Contractor having been fully paid (not just sums due in respect of the CDP).	
9	9	**Significant change:** the clause automatically applies a clause that previously had to be specifically adopted by the Parties.	The clause has been redrafted – this reflects its inclusion into the SBC/Q 2005 main contract as opposed to being a stand-alone document to be used other forms of JCT contract such as the Intermediate Form or With Contractor's Design Form. The reference in the Contract Particulars states that the default position in the Contract Particulars is that the Professional Indemnity Insurance will be for the stated value in aggregate (rather than for each claim arising out of an event) unless a contrary position is stated. State the basis on which Professional Indemnity Insurance is required.	
10	11	No change: the clause repeats the JCT 2001 Funder Collateral Warranty.		
11	12	No change: the clause repeats the JCT 2001 Funder Collateral Warranty.		
12	13	No change: the clause repeats the JCT 2001 Funder Collateral Warranty.	The limitation period in respect of the Purchaser and Tenant rights starts to count down from practical completion. Note that in respect of a section of the Works this may be some time well before the completion of the whole of the Works. The time periods follow the mode by which the SBC/Q 2005 contract itself has been executed.	

JCT 2005 Standard Building Contract With Quantities (SBC/Q 2005)

SBC/Q 2005 clause and reference title	JCT 1998 clause and reference title	Summary of change	Consequence of change and comments	New action required
13	14	No change: the clause repeats the JCT 2001 Funder Collateral Warranty.		
14		**Major change:** the amendment changes a fundamental process or procedure of the contract compared to JCT 1998.	Paragraph 14.2 had no predecessor in the 2001 Funder's Collateral Warranty, but the clause states the position that would have applied under the contract. The clause concerns the dispute resolution process following the Funder's activation of his rights. The dispute resolution process stated in the Articles will apply, thus adjudication will apply. Arbitration might apply if the Parties have adopted it in the Articles.	
Schedule 6: Forms of Bonds				
Part 1: Advance Payment Bond	Advance Payment Bond	No change: the clause repeats the text of the JCT 1998.		
Part 2: Bond in respect of payment for off-site materials and/or goods	Bond in respect of payment for off-site materials and/or goods	No change: the clause repeats the text of the JCT 1998.		
Part 3: Retention Bond	Annex 3 Bond in lieu of retention	No change: the clause repeats the text of the JCT 1998.	The Retention Bond is identical save the fact that 'termination' is referred to rather than 'determination' and that numbers are now used in preference to letters and roman numerals – clearly this has no impact on the effect of the Bond.	

Schedule 7: Fluctuations				
Part 1: Fluctuations Option A (Contribution, Levy and Fluctuations)	Clause 38	Slight amendment: the clause repeats the text of JCT 1998 but makes use of new defined terms and makes slight changes, but the effect is unchanged.	The text now refers directly to the Contract Sum rather than 'the prices contained in the Contract Sum and the Contract Sum Analysis' as it did in JCT 1998, however this change is not thought to have consequence. The term 'Base Date' is defined by reference to the Contract Particulars. The only other change is the updating of references to legislation; the provisions of the Fluctuations Option A are the same as clause 38 of JCT 1998. The relevant change is that A.1.2 now refers to the *Industrial Training Act* 1982 rather than the *Industrial Training Act* 1964.	
Part 2: Fluctuations Option B (Labour and materials cost and tax fluctuations)	Clause 39	Slight amendment: the clause repeats the text of JCT 1998 but makes use of new defined terms and makes slight changes, but the effect is unchanged.	The text now refers directly to the Contract Sum rather than 'the prices contained in the Contract Sum and the Contract Sum Analysis' as it did for JCT 1998, however this change is not thought to have consequence. The term 'Base Date' is defined by reference to the Contract Particulars. The only other change is the updating of references to legislation; the provisions of the Fluctuations Option B are the same as clause 39 of JCT 1998. The relevant change is that B.2.2 now refers to the *Industrial Training Act* 1982 rather than the *Industrial Training Act* 1964.	
Part 3: Fluctuations Option C (formula adjustment)	Clause 40	Slight amendment: the clause repeats the text of JCT 1998 but makes use of new defined terms and makes slight changes, but the effect is unchanged.		

JCT 2005 Standard Building Contract With Quantities (SBC/Q 2005)

Table of destinations: JCT (Private With Quantities) 1998 and SBC/Q 2005

This table of destinations follows the structure of the old JCT 1998 and states the new clause reference in the 2005 form.

PW 1998	SBC/Q 2005	PW 1998	SBC/Q 2005
Articles of Agreement	Articles of Agreement	1.5	3.6
First recital	First Recital	1.6	3.26
Second recital	Second Recital	1.7	1.7
Third recital	Third Recital	1.8	1.5
Fourth recital	Fourth Recital	1.9	3.3
Fifth recital		1.10	1.12
Sixth recital	Fifth Recital	1.11	1.8
Seventh recital		1.12	1.6
Article 1	Article 1	2.1	2.1 and 2.3.3
Article 2	Article 2	2.2.1	1.3
Article 3	Articles 3 and 3.5	2.2.2	2.13 and 2.14
Article 4	Articles 4 and 3.5	2.3	2.15
Article 5	Article 7	2.4.1	
Article 6.1	Article 5	2.4.2	
Article 6.2	Article 6	3	4.4
Article 7A	Article 8	4.1.1	3.10
Article 7B	Article 9	4.1.1.1	3.10.1
Attestation	Attestation	4.1.1.2	3.10.2
1.1	1.2	4.1.2	3.11
1.2	1.3	4.2	3.13
1.3	1.1 and 6.8	4.3	3.12
1.4		5.1	2.8.1

PW 1998	SBC/Q 2005
5.2	2.8.2
5.3	2.9
5.4.1	2.11
5.4.2	2.12
5.5	2.8.3
5.6	
5.7	2.8.4
5.8	1.9
5.9	
6.1.1	1.1 and 2.1
6.1.2	2.17.1
6.1.3	2.17.2
6.1.4	2.18
6.1.5	2.17.3
6.1.6	
6.1.7	
6.2	2.21
6.3	3.7.3
6A.1	3.25
6A.2	3.25
6A.3	3.26
6A.4	3.25
7	2.10
8.1.1	2.3.1
8.1.2	2.3.2
8.1.3	2.1

PW 1998	SBC/Q 2005
8.1.4	2.3.5
8.1.5	2.3
8.2.1	2.3.4
8.2.2	3.20
8.3	3.17
8.4	3.18
8.5	3.19
8.6	3.21
9.1	2.22
9.2	2.23
10	3.2
11	3.1
12	3.4
13.1.1	5.1
13.1.2	5.1
13.1.3	
13.2.1	3.14.1
13.2.2	3.14.2
13.2.3	5.2 and 5.3
13.2.4	3.14.4
13.2.5	3.14.5
13.3	3.16
13.3.1	3.16
13.3.2	
13.4.1.1	5.2.1
13.4.1.2	

**JCT 2005 Standard Building Contract
With Quantities (SBC/Q 2005)**

PW 1998	SBC/Q 2005
13.4.2	
13.5.1	5.6
13.5.2	5.6
13.5.3	5.6
13.5.4	5.7
13.5.5	5.9
13.5.6	
13.5.7	5.10
13.6	5.4
13.7	5.5
13A	5.3 and Schedule 2
14.1	4.1
14.2	4.2
15.1	
15.2	4.6
15.3	4.6
16.1	2.24
16.2	2.25
17.1	2.30
17.2	2.38
17.3	
17.4	2.39
17.5	
18.1	2.33
18.1.1	2.34
18.1.2	2.35

PW 1998	SBC/Q 2005
18.1.3	2.36
18.1.4	2.37
19.1.1	7.1
19.1.2	7.2
19.2.1	
19.2.2	3.7
19.3	3.8
19.4	3.9
19.5.1	
19.5.2	
20.1	6.1
20.2	6.2
20.3	6.3
21.1.1	6.4
21.1.2	
21.1.3	
21.2	6.5
21.3	6.6
22.1	6.7 and Schedule 3
22.2	6.8
22.3.1	6.9
22.3.2	
22A.1	Schedule 3, Option A.1 and 6.7
22A.2	Schedule 3, Option A.2
22A.3	Schedule 3, Option A.3
22A.4	Schedule 3, Option A.4

PW 1998	SBC/Q 2005
22A.5	6.10
22B.1	
22B.2	Schedule 3, Options B.1 and B.2
22B.3	Schedule 3, Option B.3
22B.4	6.10
22C.1	Schedule 3, Option C.1
22C.2	Schedule 3, Option C.2
22C.3	Schedule 3, Option C.3
22C.4	Schedule 3, Option C.4
22C.5	6.10
22D	
22FC.1	6.13
22FC.2	6.14
22FC.3	6.15
22FC.4	
22FC.5	6.16
23.1.1	2.4
23.1.2	2.5
23.2	3.15
23.3.1	2.4
23.3.2	2.6
23.3.3	2.6
24.1	2.31
24.2	2.32
25.1	2.26
25.2	2.27

PW 1998	SBC/Q 2005
25.3.1	2.28.1, 2.28.2 and 2.28.3
25.3.2	2.28.1 and 2.28.4
25.3.3	2.28.1 and 2.28.5
25.3.4	2.28.6
25.3.5	
25.3.6	
25.4	2.29
26.1	4.23
26.2	4.24
26.3	
26.4	
26.5	4.25
26.6	4.26
27.1	8.2
27.2.1	8.4
27.2.2	8.4
27.2.3	8.4
27.2.4	8.2
27.3.1	8.1
27.3.2	8.5
27.3.3	
27.3.4	
27.4	8.6
27.5	
27.6	8.7
27.7	8.8

**JCT 2005 Standard Building Contract
With Quantities (SBC/Q 2005)**

PW 1998	SBC/Q 2005
27.8	8.3
28.1	
28.2	8.9
28.3	8.10
28.4	8.12
28.5	8.3
28A.1	8.11
28A.2	8.12
28A.3	8.12
28A.4	8.12
28A.5	8.12
28A.6	8.12
28A.7	
29	2.7
30A	4.7
30.1.1.1	4.9 and 4.13
30.1.1.2	4.13
30.1.1.3	4.13
30.1.1.4	4.13
30.1.1.5	4.13
30.1.1.6	4.8
30.1.2.1	4.11
30.1.2.2	4.12
30.1.3	4.9
30.1.4	4.14
30.2	4.10 and 4.16

PW 1998	SBC/Q 2005
30.3	4.17
30.4	4.20
30.4A	4.19
30.5	4.18
30.6.1	4.5
30.6.2	4.3
30.7	
30.8	4.15
30.9	1.10
30.10	1.11
31	4.7
32	
33	
34.1	3.22
34.2	3.23
34.3	3.24
35	
36	
37.1	4.21
37.2	
37.3	4.22
38	Schedule 7, Part 1
39	Schedule 7, Part 2
40	Schedule 7, Part 3
41A	9.2
41B.1	9.4

PW 1998	SBC/Q 2005
41B.2	9.5
41B.3	9.6
41B.4	9.7
41B.5	9.8

PW 1998	SBC/Q 2005
41B.6	9.3
41C	
42	
Code of practice (clause 8.4.4)	Schedule 4

**JCT 2005 Standard Building Contract
With Quantities (SBC/Q 2005)**

3   JCT Design and Build Contract 2005

Introduction

The JCT Design and Build Contract 2005 (DB 2005) is the latest version of the JCT Standard Form of Building Contract With Contractor's Design 1998 (WCD 1998). In general terms, the 2005 version is an improved and slimmed down version of the 1998 form that has been updated and reorganised without fundamentally changing the balance of risk. The drafting and layout of the terms and conditions have been improved and a number of existing processes have been sharpened. For example, the contract now requires a far higher degree of detail from the Employer when responding to Contractors' notices for extensions of time. Terms frequently seen in schedules of amendments produced for specific projects have also been adopted. For example, if the Employer provides the details in the Articles, then the Contractor is now required to insure his design liability. In addition, certain new processes have been added, most notable of which is a Contractor's Design Submission Procedure that has been adapted from the Major Projects Form of 2003 – whilst this procedure does give the Employer a degree of scrutiny and does have a defined end point, it does not interfere with the Contractor's design responsibility or liability.

Purpose and structure of this chapter

The first part of this DB 2005 chapter will introduce the headline concepts; the second part comprises a tabular format that considers DB 2005 clause by clause with the aim of identifying and explaining the differences between WCD 1998 and DB 2005. This is a practitioners text for users familiar with the existing WCD 1998 form of contract and the exercise purposefully avoids analysing clauses that have not been changed in the 2005 version of the contract.

The structure of the table

The second part of this chapter has been set out in a tabular format to aid use as a point of reference when Contract Documents are being prepared and when the contract is used in practice on site. Reading from left to right the columns are:

- Column One: states the number and name of the DB 2005 clause;
- Column Two: states the number and name (where there is one) of the WCD 1998 clause;
- Column Three: a brief statement summarising any change and its nature. Standard descriptions are used to indicate the degree of change at a glance;
- Column Four: examines the specific significance of any changes including a consideration of the consequences of the differences for the operation of the contract; as such, this column is the heart of the book;
- Column Five: states new actions that the Parties must perform and, where relevant, provides guidance on what this means in practice.

Base documents

In this chapter DB 2005 has been compared to JCT With Contractor's Design 1998 incorporating Amendments 1: 1999, 2: 2000, 3: 2001, 4: 2002, 5: 2003 (this contract is referred to as 'WCD 1998' throughout the text).

In addition, comparison is made between DB 2005 and the following documents:

- Sectional Completion Supplement with Quantities 1998 (revised October 2003);

JCT 2005 Design and Build Contract
(DB 2005)

- Contractor's Designed Portion Supplement with Quantities 1998 (revised November 2003);
- Collateral Warranty from Contractor to Purchasers & Tenants (MCWa/P&T) 2001 Edition;
- Collateral Warranty from Contractor to Funder (MCWa/F) 2001 Edition.

Summary of the changes in DB 2005

The 2005 edition of the JCT's Contractor Design and Build form of contract is an improvement upon its predecessor and represents a steady development of the form rather than any kind of quantum leap – this is a good thing, as the form has proved effective and is popular and well-understood, thus there has been little call for a general recasting of its obligations.

The changes fall into two broad categories and the changes are considered under these headings. The first category is what might be called 'JCT 2005 changes' that introduce principles and changes found across the other JCT contracts being introduced in 2005 – many of these are structural and reflect the new layout and improved drafting style, but they also include significant substantive changes being applied across the range of JCT contracts. Such changes include the wholesale revision to the dispute resolution regime and the reduction of the Retention percentage from the previous rate of 5% to the new rate of 3% for the contracts for larger projects.

The second category comprises changes specific to DB 2005 that address the particular shortcomings or ambiguities that caused difficulty when using the WCD 1998 form in practice.

JCT 2005 changes

Changes to drafting and organisation

The usability and comprehensibility of the document as a whole has been aided by the adoption of the JCT's new regime for clause layout and drafting. Thus the 'structural' changes (as opposed to substantive changes) are found firstly in terms of the arrangement of the contract into a series of sections which group related clauses; and secondly in terms of the rewording and redrafting of familiar clauses in something approaching plain English. The division into sections has been applied in a manner that is fairly consistent with the layout and the grouping of the clauses in the other JCT contracts being released in 2005 such as the Standard Building Contract (SBC 2005) but the clauses are not exactly the same and the differences that exist are necessary to cater to the specific needs of a design and build method of procurement.

Use of titles and subheadings
The order and arrangement of the clauses of WCD 1998 was rather difficult to follow, and although the contract made some use of headings and footnotes, the 'running order' of the clauses seemed a touch arbitrary, with some clauses standing very much on their own surrounded by provisions dealing with other matters. The 2005 version of the form has grouped related clauses together in a more complete and consistent manner under one of nine section headings, and the clauses are all given titles.

Even though this means that familiar clauses are relocated, the reordering of the clauses and the use of sections and clause headings throughout greatly assist when trying to locate clauses in the text. This is certainly an improvement on WCD 1998 where one needed to refer to the small text set to one side of the main clauses if one was searching for a particular subject.

Changes to terminology

The 2005 edition of the JCT forms has seen a change to certain familiar terms or concepts and seen certain familiar but unnamed processes being given names or being captured in defined terms.

A number of familiar communications under the contract are now given names such as the 'Non-Completion Notice' (clause 2.28) and 'Practical Completion Statement' (clause 2.27). Other notable changes update the language, for example the contract is now 'terminated' rather than 'determined'. Some of the changes simply reorder already familiar names; the new term 'Notice of Completion of Making Good' (clause 2.36) replaces the WCD 1998 term 'Certificate of Completion of Making Good Defects'.

There is now an 'Adjustment to the Completion Date' rather than an 'Extension of Time', which would appear to underline the point that the time for a task can be reduced as well as extended if an element of work is removed from the Contractor.

The statement of a default position

In certain key instances WCD 2005 has moved away from relying on Parties to fill in the blanks, and the contract provides a default position from which the Parties must specifically deviate. Previous JCT contracts placed greater onus for selecting an option or incorporating a supplement upon the Parties and their advisors. There are now significant points on which a default position is stated. Users should note, however, that this is not the case for all of the choices available in the contract and a number of options must still be specifically selected.

Providing a default position will not give all Parties what they require for all projects, but it does give a degree of certainty. The stating of the default is usually done in a way that does not prevent a project-specific requirement being stated in preference or an amendment being inserted where necessary. The statement of a default position is certainly likely to assist in those cases (which are not particularly rare in construction) where the contract ends up unexecuted but the Parties have adopted its terms.

In conclusion, this is a pragmatic change that will be of great use in practice. The default position is a base position that one anticipates the JCT believes to be fair following its consideration of law, practice and commercial reality, but it comes with the acknowledgement that particular projects and sophisticated users may have specific requirements.

Changes of substance

Payment process

The provisions for the issue of payment notices and withholding notices have been more clearly stated and the clauses have been amended in a manner consistent with the other JCT 2005 contracts, thus the contract follows the *Housing Grants, Construction and Regeneration* Act 1996 ('the Construction Act') and provides for two notices to be issued to the Contractor after the interim certificate has been issued by the Architect/Contract Administrator (A/CA) – one payment notice (clause 4.10.3) and one withholding notice (clause 4.10.4). If the payment notice under clause 4.10.3 is stated in detail and with care (in terms of the statements of the sums to be withheld, with reasons for the withholding), then that payment notice may also serve as the withholding certificate with the effect that no subsequent notice is necessary. Having served a payment notice, the Employer may then serve a withholding notice, but he may serve this notice even if there has been no payment notice under clause 4.10.4.

In addition to these changes that reflect the JCT's general tidy-up of its payment clauses, there is also a specific change (dealt with below) that establishes the sum that is due to the Contractor in the event of non-payment.

Sums to be paid
The contract provides for Retention but does so at a lower level – the rate of Retention has been decreased from 5% to 3%.

Contract Particulars
This contract seeks to place all of the project-specific information in one place and in so doing it is consistent with many of the other forms that have been released in 2005. The information is gathered at the beginning of the contract in a new section called the 'Contract Particulars'. Although a large amount of this information was stated at the end of WCD 1998 in Appendix 1, certain details were spread through the Recitals, Articles and the main text of the contract.

Grouping the project-specific information should assist in document preparation and as such should improve the usability of the new form.

Dispute resolution
There have been revisions to the dispute resolution process at each level. Most significantly, DB 2005 has changed the default position from arbitration to litigation. Previous JCT forms have sought to provide for the arbitration of all disputes under the contract. The effect was such that cases before the court would be stayed pending the outcome of arbitral proceedings.

DB 2005 does not provide for arbitration of any disputes arising under the contract unless the Parties specifically select it in the Contract Particulars. This means that the Parties may use the courts to resolve their disputes. Parties are not compelled to apply the contract's adjudication clauses before issuing proceedings, but neither does issuing proceedings prevent a reference to adjudication.

The adjudication process has itself been changed substantially and will now be carried out under the rules of the statutory *Scheme for Construction Contracts (England and Wales) Regulations* 1998 ('the Scheme') – the JCT's own rules will not apply to adjudication under DB 2005.

The contract also has a new clause that suggests that the Parties consider the mediation of disputes (a process that has found much favour in the courts).

Third party rights
The construction contract between Employer and Contractor is only the most obvious of the obligations a construction project puts in place. Users will be aware that the English law of contract only enforces contract obligation on the Parties to the contract and that a practice developed to accommodate the needs of others that might be interested in a project whereby sub-contractors, designers and Contract Administrators entered into additional contracts with Funders and Purchasers (usually described as 'collateral warranties'). The provisions of the *Contracts (Rights of Third Parties) Act* 1999 sought to redraw some of the rules of privity of contract, but as they were not mandatory they were usually excluded from construction contracts to the maximum extent possible.

WCD 1998 did not include terms for the provision of collateral warranties, nor did it provide standard forms of warranty – these were published by JCT as stand-alone supplements and had to be purchased and incorporated on a job-by-job basis. Purchasers or investors are highly likely to see design liability as critical to their ability to maintain the risks associated with any business connected with the property being constructed, and thus collateral warranties are very common for design and build projects and if the Contractor has sub-contracted the design works, then a collateral warranty is almost certainly going to be required from that designer.

Section 7 of DB 2005 makes provision for a standard set of rights for the kind of third party that typically requires them due to a specific interest in construction activity rising out of the provision of finance or the purchase of assets. Users should note that these rights are not mandatory and unless the Parties specifically select them in Part 2 of the Contract Particulars, there will be no obligation to provide such rights. The rights themselves are set out in Schedule 5 and may either take the form of collateral warranties that are executed in documentary form, or alternatively by adoption of the *Contracts (Rights of Third Parties) Act* 1999.

The third party rights set out in DB 2005 provide a reasonable balance of the requirements of third parties with the obligations that can be accepted by Contractors, professionals and sub-contractors.

The contract provides a standard form set of rights for a Funder; these apply throughout the construction period. It also provides for step-in rights for the Funder that enable the Funder to take the place of the Employer and operate the construction contract, although they no longer contain a net contribution clause in favour of a Funder. The Purchaser and Tenant (P&T) rights are more limited than those of the Funder and only apply from practical completion onwards, whereas Funder rights may apply to limit the conduct of a Contractor where he would otherwise be free to act without reference to the concerns of any third party (including P&T).

In addition to the rights that the P&T or Funder may have against the Contractor, provision is made for collateral warranties to be provided by sub-contractors and professionals to third parties. The terms are substantially similar to existing JCT documents that had been previously published as supplements.

The incorporation of the JCT standard warranties and the obligations to ensure that they are provided is certainly welcome, although this is tempered by the reality that Funders are uniquely placed to ensure that their particular requirements are met and therefore may well insist on their own conditions and/or amendments.

Termination
The termination clause has been updated and the insolvency definitions changed – the most significant change is that the contract will not immediately terminate in the event of insolvency. Under DB 2005 a notice of termination must be served to bring a contract to an end due to the insolvency of a Party, whereas WCD 1998 was automatically determined upon the occurrence of most insolvency events.

Electronic communications
The Parties may agree that certain communication required by the terms of the contract can be sent by electronic means. If this is required, they must complete the section in the Contract Particulars. This amendment replaces the WCD 1998 scheme for an Electronic Data Interchange and has a number of implications. Of these, the most significant is that, if correctly drafted, the contract might provide that all communication and notice required under the contract could be sent electronically.

Many projects are, to all intents and purposes, managed in this way, with instructions and payment certificates being issued by email. It is often the case that terms and conditions require hard copy notices to be issued and thus the electronic or faxed document arrives some days before the written notice finally arrives.

The main risk of such change is that it downgrades the apparent significance of documents that have very serious consequences if they are not dealt with in an appropriate manner. Another risk is the

111

perennial problem where a difficulty that has been elliptically alluded to in an email between staff is later said to represent good notice. It would seem that professionalism would safeguard against such consequences. Provided that the staff drafting the correspondence do it correctly and state its purpose clearly, and that the staff receiving the correspondence are alive to the potential that each email may have, then quick and clear communication can ensure that the contract is operated more effectively.

Mandatory bonds

The contract provides additional security for Employers providing an advance payment to the Contractor and adopts the default position that the Contractor is obliged to provide an Advance Payment Bond in respect of the sums advanced. The form of the bond is appended to Schedule 6 of the Contract. The provision that advance payment ought to be so secured may lead to an increase in the use of advance payments.

Changing the time for completion

The adjustment to the time for completion retains the integrity of the construction period despite events that may be the fault of the Employer, and ensures that the Contractor will still be liable to pay liquidated damaged if he fails to achieve practical completion on time.

Clause 2.25.3 of DB 2005 introduces a major change to the information that the Employer must provide when responding to the Contractor's notice of a Relevant Event. The Employer must now state the reasons that he takes into account when adjusting the Completion Date – in this respect the new contract almost follows the withholding notice format.

The clause provides that the Employer must respond to the Contractor's notice by identifying the acceptable Relevant Event and stating the period of time due against it. Users, particularly Employer's Agents, should be careful to update their processes and ensure that this new provision is satisfied when adjusting the Completion Date under DB 2005.

The Relevant Events that may give rise to a need to adjust the Completion Date have been slightly updated for DB 2005. The Contractor is now more thoroughly responsible for his own supply chain (in terms of both labour and materials). This also reflects present understandings of the service provided by Contractors and their bargaining power, as well as any wider economic significance that could be interpreted.

Changes specific to the Design and Build form

Sums due as Interim Payments when payment process is not complied with

Clause 30.3.5 of WCD 1998 provided that the sum stated in a Contractor's Application for Interim Payment would be the sum due to the Contractor in the event of the Employer failing to pay the sum set out in the Application and yet also failing to comply with the payment provisions by not issuing a payment notice under clause 30.3.3 or a withholding notice under clause 30.3.4.

The arguments in such circumstances were inevitably fraught – concerning the correctness of the various notices in terms of form and content (application, payment notice, withholding notice) and also in terms of whether the sums claimed were indeed due under the contract.

This payment pitfall is a current issue for any project proceeding with a WCD 1998. At the time of writing, there is no English law authority on the point. Some Adjudicators are minded to consider the question of the amount due when deciding what, if anything, ought to be paid; others consider that if the payment process had not been applied they were bound to rubber-stamp a mandate to pay the Contractor's Application for Interim Payment.

Clause 4.10.5 of DB 2005 deals with the point in a different way, providing that if the notices are not issued correctly and yet sums are withheld, then the sum due to the Contractor is determined under clause 4.8. Clause 4.8 states the basis on which an Interim Payment is due, thus in the event of the Employer illegitimately withholding sums, the Contractor is only ever entitled to be paid the amount due rather than such sum as he states in an Application for Interim Payment.

Insurance for design liability

The design and build contract places a design obligation upon the Contractor, but WCD 1998 did not state the requirement for the Contractor to take out and maintain specific insurance in this regard. This position has been addressed in DB 2005 with an express requirement for the Contractor to take out and maintain professional indemnity insurance under clause 6.11. The requirement for such insurance is an option and unless the Contract Particulars state the requirement *and* specifically state the minimum level that the insurance is to be maintained at, then the Contractor is not obliged to provide such insurance at all. This point reinforces the general suggestion that great care must be taken to complete the Contract Particulars correctly. The Particulars provide the option for the Parties to select the insurance on either an 'each and every' or 'aggregate' basis and defaults to an aggregate basis unless the Parties make alternative provision.

The obligation to provide the insurance is not absolute, as the clause makes provision for the exceptional circumstances when the insurance is not available at 'commercially reasonable rates'. Clause 6.22.3

appears to be a get-out for the Contractor but it is not stated in objective terms as to refer to 'market rates'. The purpose of the clause may be to assist a Contractor who is obliged to maintain such insurance against a background of an industry-wide escalation in premiums, but this belies the fact that premiums also take into account other factors, such as the performance of the person seeking insurance. Thus a Contractor whose own track record has led to it being offered insurance only at rates it considers to be 'commercially unreasonable' may well seek to escape clause 6.11.1 and its obligation to insure. In many cases these may well be the designers for whom such insurance is most essential.

Cost of design activity

A variation may have the consequence of increasing or decreasing the amount of design work that is required of the Contractor. The need for extra design work and drawings, or the requirement for less, may represent either a significant saving or indeed additional cost to the Contractor – particularly if the work is carried out by a sub-contractor. Clause 12 of WCD 1998 made express provision for costs associated with design and thus did not enable to the Employer to secure the benefit of such saving and did not expressly deal with additional or reduced cost of design as a result of addition or omission of design work in the Valuation Rules. This is now dealt with in the Valuation Rules (found at clause 5.4 of DB 2005) that apply if the Parties are unable to agree the sum that is to be paid in respect of a variation. Clause 5.4.1 states that 'Allowance shall be made in such Valuations for the addition or omission of the relevant design work'. As a result of this change, the cost consequences of varying the Works will be to the account of the Employer.

Contractor's Design Submission Procedure

The Procedure has been adopted across those of the new forms that incorporate the provisions that were formerly stated in the separate

'Contractor's Design Portion Supplement', but is of particular application to this form of contract and addresses several of the shortcomings of the WCD 1998 provisions for changes affecting design. The Contractor's Design Submission Procedure is found in Schedule 1 towards the end of the contract.

The new procedure is mandatory and requires all Design Documents be issued to the Employer before they are used. The Employer has 14 days to consider the drawings, and if he fails to act the Schedule makes express provision that the Contractor may proceed using the drawings submitted. Thus the Contractor's course of action is clear if the Employer does not respond in the time-period stated in the Design Submission Procedure.

Users should not mistake this process for a full vetting or approval process – Contractors in particular should note that clause 8.3 of the Schedule provides that:

'… neither compliance with the design submission procedure in this Schedule nor with the Employer's comments shall diminish the Contractor's obligations to ensure that the Contractor's Design Documents and Works are in accordance with this Contract'.

The main benefit of the Contractor's Design Submission Procedure is that it gives a certain outcome – this is also its main drawback, as it is a process that will require swift attention by the Employer and his team.

The novated design team

It is perhaps worth noting the practice that has developed whereby the Employer identifies a design team and requires that team to develop the design with the explicit intention that the same team will then become the Contractor's team once the Contractor is appointed under a Design and Build form of contract.

The transfer of the design team from Employer to Contractor is usually carried out by some form of novation. Neither the DB 2005 form nor any of its predecessors have addressed this thoroughly established practice by providing a mechanism in the main contract, or by creating a set of professional appointments that take account of this inevitable transfer. The result is that such transfers frequently require bespoke documentation and negotiation – indeed, it is open to question whether such transfers can be considered 'novations' if the terms of the contract are amended in this process.

In addition to such questions, there is some academic comment to the effect that the technique of transferring the same design team should not be considered true design and build and that it does not capture all of the benefits that the design and build method of procurement can achieve. Certainly, as with most procurement techniques, there are advantages and disadvantages to this method of working, however given the frequency with which the technique is employed, the merits must be visible to many. In consequence, it is worth noting that DB 2005 is robust enough to accommodate such techniques; indeed, the new Contractor's Design Submission Procedure is likely to ensure that the consequence of 'design development' after commencement is likely to be managed with greater visibility and clarity of outcome.

Whilst the 2005 version of the DB form is an improvement on its predecessor in the case of novation of design teams, it seems that we will have to wait for future editions of the design and build form to close the gap that exists between theory and practice and provide a standard mechanism for transferring a design team from the Employer to the Contractor. It does seem to be something of an anomaly for the standard form to ignore what can be one of the most difficult transactions in design and build procurement. At the very least, if this cannot be incorporated into the standard form it might be published separately as a supplement.

Design and build projects in practice

In many respects 'design and build' is simply a procurement technique and in other sectors of industry this method of procurement is very much the norm. However, the traditional forms of construction contract where the Employer's team is responsible for all design elements (exemplified by JCT's Standard Form of Building Contract) are completely incompatible with this technique of procurement – consequently a specific form of contract is required.

In practice, the design and build forms of JCT contract have been enthusiastically embraced by both Employers and Contractors, and the popularity of this method of procurement appears to be increasing with certain Contractors specialising in this type of work alone.

In addition to Employers and Contractors, independent construction professionals have also responded to the opportunities and challenges associated with the design and build formula. This is despite the fact that the opportunities for the traditional construction professionals are perhaps less obvious, as they are not identified and given explicit roles. Thus the JCT Design and Build forms do not create roles or responsibilities for an 'Architect', 'quantity surveyor', 'project manager' or even 'contract administrator'. Ultimately, many of the same services are still essential for a project to be delivered successfully – given the large scale of many design and build projects these services are usually critical, but the difference may be that a professional designer is an employee of the Contractor or alternatively is a sub-contractor to him.

115

JCT 2005 Design and Build Contract (DB 2005)

DB 2005 clause and reference title	WCD 1998 clause and reference title	Summary of change	Consequence of change and comments	New action required
Articles of Agreement	Articles of Agreement		*Generally:* Although the Articles have been substantially derived from WCD 1998, they have been generally sharpened and the words used are more contemporary. There is now a space for Parties to state their company number on page 1 of the contract where the Parties are identified. Stating the company number gives additional clarity as to the identity of the Parties, as there is a risk that the company named is not the company whose address is stated. The company number provides an additional reference point that can easily be checked (for example, by research at Companies House or the Companies House website: www.companieshouse.co.uk), and is also of value when considering a company's financial position. Note also that the 'health warning' concerning the suitability of the use of Contractor Design and Build forms as opposed to a Standard JCT 1998 contract With a Contractor's Designed Portion has been deleted.	Companies incorporated under the *Companies Act* should state their company number.
Recitals				
First Recital *(the Works)*	First Recital	Clause redrafted: the clause has been significantly reworded and gives additional clarity but does not give rise to new obligations.	The text that remains is similar to WCD 1998 and has the same effect. The significance lies in what has been deleted: the Recital no longer refers to the	

			Construction (Design and Management) Regulations 1994 ('the CDM Regulations').	
Second Recital *(Contractor's Proposals and Contract Sum Analysis)*	Second Recital	Clause redrafted: the clause has been significantly reworded and gives additional clarity but does not give rise to new obligations.	The clause has been substantially redrafted and has gained some clarity as a result. References to the Contract Sum have been relocated to the Articles and the second paragraph of the old clause (dealing with CDM compliance) has been deleted. The provisions relating to the Contractor's role as Planning Supervisor have been relocated to the Contract Particulars.	
Third Recital *(Employer's satisfaction with Contractor's Proposals)*	Third Recital	Slight amendment: the clause repeats the text of WCD 1998 with slight changes that improve the drafting or syntax, but the effect is unchanged.		
Fourth Recital *(the Construction Industry Scheme)*	Fourth Recital	Slight amendment: the clause repeats the text of WCD 1998 with slight changes that improve the drafting or syntax, but the effect is unchanged.		
Fifth Recital *(Sections)*	Article 8	Clause redrafted: the clause has been significantly reworded, but does not give rise to new obligations.	A greatly simplified provision gives the Parties the option to divide the Works into Sections. The information should be stated in the Contract Particulars.	
The Articles				
Article 1: Contractor's Obligations	Article 1	Clause redrafted: the clause has been significantly reworded and gives additional clarity but does not give rise to new obligations.	The obligation to 'complete the design for the Works and carry out and complete the construction of the Works' is the same as that under WCD 1998 but is improved by the direct reference to this activity being 'in accordance with the Contract Documents'. The reference to the obligation being undertaken for a consideration has been removed, but is in any event unnecessary.	

117

DB 2005 clause and reference title	WCD 1998 clause and reference title	Summary of change	Consequence of change and comments	New action required
Article 2: Contract Sum	Article 2	**Significant change:** the clause repeats most of the text of the WCD 1998, but it is also amended in a way that does affect the operation of the clause.	The expression that the Contract Sum is 'VAT-exclusive' is a new addition and states the position as it is generally understood to be.	
Article 3: Employer's Agent	Article 3	Clause redrafted: the clause has been significantly reworded and gives additional clarity, but does not give rise to new obligations.	The redraft is a less restrictive statement of the appointment of the Employer's Agent. Under DB 2005 the appointment is 'for the purposes of this Contract', whereas under WCD 1998 the appointment was of 'the Employer's Agent as referred to in clauses 5.4 and 11', both of which were clauses of fairly limited scope. The items for which the Employer's Agent may issue or receive correspondence with contractual significance has not changed save that the drafting is, again slightly looser. Specifically, Article 3 of WCD 1998 states that the Employer's Agent may give and/or receive notice (etc.) or otherwise act under any other of the Conditions. The difference between the forms as a result of the removal of the word 'other' in DB 2005, which makes the range of the Employer's Agent slightly wider as it empowers the Contract Administrator to give or receive the notice 'and otherwise act for the Employer under any of the Conditions'. For the avoidance of any doubt, whilst there is a slight difference, it is of very limited practical application.	
Article 4: Employer's Requirements and Contractor's Proposals	Article 4	**Significant change:** the clause has been redrafted and this affects its operation.	Although largely the same, the new version of Article 4 does not require the documents to be signed by the Parties.	

			It is unclear why the standard of certainty has been lowered in this instance and it runs counter to the trend of the other JCT 2005 contracts where the requirements for initialling and signing documents has increased rather than decreased.	
Article 5: Planning Supervisor	Article 7.1	Slight amendment: the clause repeats the text of WCD 1998 with slight changes that improve the drafting or syntax, but the effect is unchanged.		
Article 6: Principal Contractor	Article 7.2	Slight amendment: the clause repeats the text of WCD 1998 with slight changes that improve the drafting or syntax, but the effect is unchanged.	DB 2005 provides a space for the name of the Principal Contractor in the event that it is not to be the Contractor.	
Article 7: Adjudication	Article 5	Slight amendment: the clause repeats the text of WCD 1998 with slight changes that improve the drafting or syntax, but the effect is unchanged.		
Article 8: Arbitration	Article 6A	**Major change:** the amendment changes a fundamental process or procedure of the contract compared to WCD 1998.	The change is that arbitration is not the default dispute resolution mechanism for the DB 2005 contract. This Article is optional – it must be specifically selected in the Contract Particulars to have effect. Note that if the Parties make the necessary change to the Contract Particulars, then the provisions of this Article are unchanged in comparison to WCD 1998.	
Article 9: Legal proceedings	Article 6B	**Major change:** the amendment changes a fundamental process or procedure of the contract compared to WCD 1998.	The change is that Parties may refer disputes or differences to the court for resolution, whereas under WCD 1998 the position had been that arbitration was the default dispute resolution method. The consequence of this under WCD 1998 had been that the courts would enforce the arbitration agreement and stay court process in deference to the *Arbitration Act* 1996 and the Parties' arbitration agreement.	

JCT 2005 Design and Build Contract (DB 2005)

DB 2005 clause and reference title	WCD 1998 clause and reference title	Summary of change	Consequence of change and comments	New action required
			Although this position is now reversed, it should be noted that neither Article 8 nor Article 9 oust, delay or impair the statutory right of a Party to a construction contract to have a dispute referred to an Adjudicator. Thus adjudication might run concurrently with arbitration, litigation or even conceivably at the point of trial.	
Contract Particulars				
Part 1 General				
Fourth Recital and clause 4.5	Appendix 1	Slight amendment: the clause repeats the text of WCD 1998 with slight changes that improve the drafting or syntax, but the effect is unchanged.		
Fifth Recital (description of Sections)	N/A	**New definition:** this clause provides a new definition of a term or of a process.	Whilst the WCD 1998 made provision for the completion of the work in Sections, it did not provide for a description of the Sections. Accordingly, this entry in the Contract Particulars is new.	If the work is to be carried out in Sections, then describe the Sections at this point in the Contract Particulars.
Article 4: Employer's Requirements	Article 4, Appendix 3	**Significant change:** the clause repeats most of the text of the WCD 1998, but it is also amended in a way that does affect the operation of the clause.	The WCD 1998 version of Article 4 and Appendix 3 contained the requirement that the Employer's Requirements be signed as well as being listed. DB 2005 does not state the requirement for the signing of these documents. The relative merits of the new method are open to debate, but there is nothing to prevent the Parties from signing the documents in addition to completing this section of the Contract Particulars.	The documents containing the Employer's Requirements are to be stated, referenced or otherwise identified at this point in the Contract Particulars.

Article 4 (Contractor's Proposals)	Article 4 and Appendix 3	**Significant change:** the clause repeats most of the text of the WCD 1998, but it is also amended in a way that does affect the operation of the clause.	The WCD 1998 version of Article 4 and Appendix 3 contained the requirement that the Contractor's Proposals be signed as well as being listed. DB 2005 does not state the requirement for the signing of these documents. The relative merits of the new method are open to debate, but there is nothing to prevent the Parties from signing the documents in addition to completing this section of the Contract Particulars.	The documents containing the Contractor's Proposals are to be stated, referenced or otherwise identified here.
Article 4 (Contract Sum Analysis)	Article 4 and Appendix 3	**Significant change:** the clause repeats most of the text of the WCD 1998, but it is also amended in a way that does affect the operation of the clause.	The WCD 1998 version of Article 4 and Appendix 3 contained the requirement that the Contract Sum Analysis be signed as well as being listed. DB 2005 does not state the requirement for the signing of these documents. The relative merits of the new method are open to debate, but there is nothing to prevent the Parties from signing the documents in addition to completing this section of the Contract Particulars.	The documents containing the Contract Sum Analysis are to be stated, referenced or otherwise identified in the Contract Particulars at this point.
Article 8 (Arbitration)	Articles 6A and 6B	**Major change:** the amendment changes a fundamental process or procedure of the contract compared to WCD 1998.	WCD 1998 contained an arbitration agreement the effect of which was that if the Parties were in dispute, they would take that dispute to an Arbitrator rather than to the courts. If proceedings are issued to litigate a dispute, then the courts would stay an action in the courts in deference to the Parties' agreement to adopt arbitration (and the *Arbitration Act 1996* that supported it). The new default position for DB 2005 is diametrically opposite to that stated in WCD 1998 and is that arbitration will not apply to disputes arising under the contract and disputes will be resolved through the courts.	If the Parties want to arbitrate disputes rather than litigate them, then they must make the necessary changes to the Contract Particulars at this point.

DB 2005 clause and reference title	WCD 1998 clause and reference title	Summary of change	Consequence of change and comments	New action required
			If the Parties want arbitration to be the dispute resolution method for the contract instead of litigation, then this must be stated in the Contract Particulars at this point. If the Parties decide that arbitration would be appropriate for a specific dispute even though they have not selected arbitration as the default mechanism, then they are free to agree to this on a case-by-case basis. Failing such agreement for a particular dispute, the courts will not stay legal proceedings in favour of arbitration. Note also that adjudication will apply to construction contracts irrespective of the selection made.	
1.1 Base Date	1.3	No change: the clause repeats the text of WCD 1998.		
1.1 Date for Completion of the Works	1.3 Appendix 1 and Appendix 1 (Sectional Completion)	No change: the clause repeats the text of WCD 1998.	WCD 1998 provided two versions of Appendix 1, one of which provided for completion in Sections and one that provided for completion to be achieved at a particular point in time.	
1.7 Address for Service	1.5	**Significant change:** the clause repeats most of the text of the WCD 1998, but it is also amended in a way that does affect the operation of the clause.	This new clause gives greater certainty than WCD 1998 as it provides a definite location for correspondence – either the address stated in the Contract Particulars or, in its absence, the address at the beginning of the Articles of Agreement. This avoids doubts that might arise when it is difficult to determine which of several offices is the *principal office* and should ensure that a Party needing to issue a notice may always be able to serve a valid notice on the other Party.	Address, telephone and fax numbers to be provided by each Party.

1.8 Electronic communications	1.8 Electronic Data Interchange	**Major change:** the amendment changes a fundamental process or procedure of the contract compared to WCD 1998.	The Contract Particulars provide a space for the Parties to make a list of the communications, documents and notices etc. that can be validly sent electronically. If the Parties don't provide such a list, then the contract default position is that communications must be in writing and the consequence would be that electronic documents would have no formal status under the contract. The JCT has not suggested which of the communications required by the contract might sensibly be transferred electronically. At one end of the sliding scale would be a notice of termination, which it is suggested would not be appropriate for electronic notice; and at the other end, the lower-level enquiries and requests for information where it would be practical. Difficulty may arise with the matters in the middle; for example, perhaps it is appropriate for withholding notices to be sent by email but inappropriate for Instructions to be sent by email. The WCD 1998 provisions supporting an 'electronic data interchange' (and the Supplemental Provisions required) are now defunct and have not been repeated in DB 2005.	Decide whether or not to consent to valid notice or communication by email, etc. If this is selected, then the specific items of correspondence and the form of communication must be set down in this section of the Contract Particulars.
2.3 Date of Possession of the Site	Appendix 1 Ref to 23.1.1	Slight amendment: the clause repeats the text of WCD 1998 with slight changes that improve the drafting or syntax, but the effect is unchanged.	Again provision is made for both sectional and non-sectional completion. This is done in a single section of the Contract Particulars, rather than a separate appendix for each method of completion (as it was set out in WCD 1998).	
2.4 and 2.26.3 Deferment of possession	Appendix 1 Ref to 23.1.2, 25.4.14, 26.1	Slight amendment: the clause repeats the text of WCD 1998 with slight changes that improve the drafting or syntax, but the effect is unchanged.	Provision is made for both sectional and non-sectional completion. This is done in a single section of the Contract Particulars, rather than a separate appendix for each method of completion (as it was set out in WCD 1998).	

DB 2005 clause and reference title	WCD 1998 clause and reference title	Summary of change	Consequence of change and comments	New action required
2.17.3 Limit of Contractor's Liability for loss	Appendix 1 Ref to 2.5.3	Clause redrafted: the clause has been significantly reworded and gives additional clarity but does not give rise to new obligations.	There is a small degree of change in that this section of the Contract Particulars states that the liability need not be capped at all. If no figure is stated in the Contract Particulars, then the liability of the Contractor would be uncapped.	
2.29.2 Liquidated Damages	Appendix 1 Ref to 24.2.1	**Significant change:** the clause repeats most of the text of the WCD 1998, but it is also amended in a way that does affect the operation of the clause.	In addition to the familiar provision for the statement of the sum in damages for non-completion, DB 2005 provides space for the Parties to provide details of the liquidated damages to be paid against late completion of Sections. WCD 1998 required the Parties to look to the Contract Sum calculate a sum in liquidated damages for each Section. The WCD 1998 method was a more complex process that happened at the time when damages were allegedly due, and this could prove divisive in practice.	
2.34 Section: Section Sums	17.1.4	**Significant change:** the clause has been significantly reworded and gives additional clarity, but does not give rise to new obligations.	This new clause assists to preserve the Employer's right to liquidated damages when the Employer has taken early possession of parts of the site where the Works are to be carried out in Sections. The statement of the Section Sums (sums for each Section) should make the calculation of the pro rata adjustment more manageable.	
2.35 Rectification Period	Appendix 1 Ref to 16.2, 17 and 30	Clause redrafted: the clause has been significantly reworded but does not give rise to new obligations.	What was the 'Defects Liability Period' for WCD 1998 is now the 'Rectification Period' for the purposes of DB 2005.	

			As with WCD 1998, the Parties may state their own Rectification Period in preference to the default six-month period. The new form also provides space for the Parties with a project to be completed in Sections to state different Rectification Periods in respect of each of the Sections.	
4.6 Advance Payment	Appendix 1 Ref 30.1.1.2	Slight amendment: the clause repeats the text of WCD 1998 with slight changes that improve the drafting or syntax, but the effect is unchanged.	The clause is the same as WCD 1998 save that the last paragraph concerning the requirement for an Advance Payment Bond is now found in the subsequent part of the Contract Particulars.	
4.6 Advance Payment Bond	Appendix 1 Ref 30.1.1.2	**Major change:** the amendment changes a fundamental process or procedure of the contract compared to WCD 1998.	The default position for DB 2005 is that a Contractor who receives an Advance Payment will also be required to provide an Advance Payment Bond. WCD 1998 did not provide a default position, thus if no preference was stated then there would be no requirement for an Advance Payment Bond.	
4.7 Method of payment – alternatives	Appendix 2 Method of Payment	Clause redrafted: the clause has been significantly reworded, but does not give rise to new obligations.	WCD 1998 provided two separate sets of appendices depending on whether or not the Works were to be completed in Sections. DB 2005 retains the same default position as WCD 1998, but states this more clearly than its predecessor.	
4.15.4 Listed Items – uniquely identified	Appendix 1 Ref to 15.2.1	No change: the clause repeats the text of WCD 1998.		
4.15.5 Listed Items – not uniquely identified	Appendix 1 Ref to 15.2.2	No change: the clause repeats the text of WCD 1998.		
4.17.1 Retention Percentage	Appendix 1 Ref to 30.4.1.1	**Major change:** the amendment changes a fundamental process or procedure of the contract compared to WCD 1998.	The default Retention Percentage is 3% for DB 2005, whereas WCD 1998 provided for a rate of 5%.	The percentage sum is different as a result of this change and thus any standard documentation should be adjusted to reflect this change.

125

DB 2005 clause and reference title	WCD 1998 clause and reference title	Summary of change	Consequence of change and comments	New action required
4.18 and Schedule 7: Fluctuations Option	35 and Appendix 1 Ref to 35	Slight amendment: the clause repeats the text of WCD 1998 with slight changes that improve the drafting or syntax, but the effect is unchanged.	WCD 1998 contained a guidance note concerning the non-adjustable element when the contract involves a local authority; this has been removed.	
6.4.1.2 Contractor's insurance – injury to persons or property	Appendix 1 Ref to 21.1.1	Slight amendment: the clause repeats the text of WCD 1998 with slight changes that improve the drafting or syntax, but the effect is unchanged.		
6.5.1 Insurance – Liability of the Employer	Appendix 1 Ref to 21.2.1	Slight amendment: the clause repeats the text of WCD 1998 with slight changes that improve the drafting or syntax, but the effect is unchanged.	The DB 2005 draft makes it clear that the Employer is at liberty to maintain insurance at a higher level that the minimum required by the contract.	
6.7 and Schedule 3 insurance of the Works	Appendix 1 Ref to 22.1	Clause redrafted: the clause has been significantly reworded and gives additional clarity but does not give rise to new obligations.		
6.7 and Schedule 3 Insurance Option A	Appendix 1 Ref to 22A, 22B.1, 22C.2	**Major change:** the rights of the Parties have been changed by this amendment compared to WCD 1998.	Unlike WCD 1998, DB 2005 provides a default position. The effect is that if the Parties do not put in their own figure, DB 2005 defaults to a cover of 15% of the value of the loss to be available to cover professional fees.	Consider whether professional fees are adequately addressed by the default cover rate of 15%.
6.7 and Schedule 3 Insurance Option A	Appendix 1 Ref to 22A.3.1	Slight amendment: the clause repeats the text of WCD 1998 with slight changes that improve the drafting or syntax, but the effect is unchanged.		
6.11 Professional indemnity insurance	N/A	**Major change:** the amendment changes a fundamental process or procedure of the contract compared to WCD 1998.	The Contractor's design liability must now be insured with a professional indemnity insurance policy. This was neither a requirement of WCD 1998, nor was it a provision of the JCT's 1998 CDP	State the value of professional indemnity insurance that the Contractor must carry in respect of design liability.

			Supplement, but was frequently a requirement added by Employers in their standard schedules of amendments and many Contractors carrying out such work maintained this insurance in any event. Note that if no sum is stated, then the Contractor will not be obliged by the contract to insure his liability. Note also that the default period for this insurance to be maintained is six years. If the agreement is to be executed as a deed, then the Parties might wish to consider selecting the 12-year period.	
6.13 Joint Fire Code	Appendix 1 Ref to 22FC.1	Slight amendment: the clause repeats the text of WCD 1998 with slight changes that improve the drafting or syntax, but the effect is unchanged.	The note found in WCD 1998 has been moved to the footnotes. That note said that the information stated in that section of the Appendix was provided by the Contractor (at his risk rather than at any risk to the Employer) – it applied when the Contractor was required to secure the All Risks Insurance under clause 22A.	
6.16 Joint Fire Code	Appendix 1 Ref to 22FC.5	**Significant change:** the clause repeats most of the text of the WCD 1998, but it is also amended in a way that does affect the operation of the clause.	If the Contract Particulars do not state a preference, then the default position will apply and the Contractor will be obliged to bear the cost of compliance with amendments or revision to the Joint Fire Code. The change is the fact that WCD 1998 did not express a default position.	
7.2 Assignment/grant of rights by Employer	Appendix 1 Ref to 18.1.2	Slight amendment: the clause repeats the text of WCD 1998 with slight changes that improve the drafting or syntax, but the effect is unchanged.		
8.9.2 Period of suspension	28.2.2	**Major change:** the amendment changes a fundamental process or procedure of the contract compared to WCD 1998.	The change is that the default period has changed in comparison to WCD 1998. The period has *increased* from a suspension of one month to a period of two months.	

DB 2005 clause and reference title	WCD 1998 clause and reference title	Summary of change	Consequence of change and comments	New action required
			The Parties may prefer to state their own period.	
8.11.1.1–8.11.1.6 Period of suspension	28A1.1.1–28A.1.1.7	**Major change:** the amendment changes a fundamental process or procedure of the contract compared to WCD 1998.	The change is that the default period has changed compared to WCD 1998. The period has *reduced* from a suspension of three months to a period of two months. The Parties may prefer to state their own period.	
9.2.1 Adjudication	Appendix 1 Ref to 39A.2	**Major change:** the amendment changes a fundamental process or procedure of the contract compared to WCD 1998.	The provisions for the appointment of an Adjudicator under DB 2005 are different to WCD 1998 in a number of respects. DB 2005 provides that the Parties may nominate a specific individual in the Contract Particulars to act as Adjudicator – this was not provided for in WCD 1998. If the Contract Particulars have not been completed, then a referring Party may select any of the listed Adjudicator Nominating Bodies (ANBs). Under WCD 1998 the contract defaulted to the Royal Institute of British Architects (RIBA) as ANB. Note also that the Chartered Institute of Arbitrators (CIA) has been added to the list of ANBs.	
9.4.1 Arbitration	Appendix 1 Ref to 39B.1	**Major change:** the amendment changes a fundamental process or procedure of the contract compared to WCD 1998.	If arbitration has been selected, the clause does provide a default Arbitrator selection process that will apply in the event that the Parties have failed to state a preference. Note, however, that for DB 2005 this is the Chartered Institute of Arbitrators (CIA), whereas under WCD 1998 the contract defaulted to the Royal Institute of British Architects (RIBA).	

			The remit of the appointer under DB 2005 is slightly wider (or at least the role is more explicit) than that under WCD 1998 in that the appointment of a *replacement* Arbitrator is also covered. As such, this is a more complete draft that provides a resolution when a specific difficulty is encountered.	
Schedule 2: Supplemental Conditions	Appendix 1 Ref to Article 1	No change: the clause repeats the text of WCD 1998.		
Schedule 2: Site Manager	N/A	**Significant change:** this is a new clause that introduces new provisions but is not sufficiently significant to be a major change.	The provision allowing the replacement of the Contractor's Person-in charge with a Site Manager is a Supplemental Condition set out in Schedule 2 of DB 2005.	
Part 2 Third Party Rights and Collateral Warranties	N/A	**Major change:** this is a wholly new clause for the WCD 2005.	The third party rights and collateral warranty regime for DB 2005 is new and has no precedent in WCD 1998. It is a series of options (set out in full in section 7) that provides that the Contractor (and in some instances a sub-contractor) accepts certain obligations to persons other than the Employer. In the past, Parties have tended to adopt the separate supplemental collateral warranties published by the JCT or draft their own pro forma warranties in order to provide rights to third parties. DB 2005 presents an additional method of providing rights to third parties that is an alternative to the provision of collateral warranties. Since its introduction many Parties have tended to exclude the application of the *Contracts (Rights of Third Parties) Act* 1999 ('the Third Party	Users must decide if they want to provide third party rights and who should benefit from them. If third party rights are to be provided, then this must be set out in detail in Part 2 of the Contract Particulars. Unless there is a specific statement as to the rights that are to apply, then there will be no third party rights or collateral warranties under the contract.

DB 2005 clause and reference title	WCD 1998 clause and reference title	Summary of change	Consequence of change and comments	New action required
			Rights Act'). DB 2005 provides that Parties may adopt the Third Party Rights Act to support a schedule of third party rights and thereby obviate the need for the production and execution of warranty documents. The actual rights themselves are set out in clause 7 and Schedule 5 of the contract. Note that the default position is that there are to be no rights under the Contract for Purchasers/Tenants or Funders. Thus it is crucial that if third party rights are required this is indicated at this point in the Contract Particulars.	
Attestation	Attestation pages headed 'As Witness the hands …'	**Significant change:** the clause repeats most of the text of the WCD 1998, but it is also amended in a way that does affect the operation of the clause.	The attestation process is unchanged, but the words used are more modern and the 'Note on Execution' in DB 2005 is improved and is clearer than that in WCD 1998. In simple terms, DB 2005 has a much less cluttered execution page. There is now space for witnesses to state their name and address in addition to their signature – perhaps this will result in some level of increased verity of the execution of the documents. Like WCD 1998, DB 2005 provides attestation forms for limited companies and individuals. The guidance note points out that the attestation may not be correct for housing associations, partnerships and foreign companies, but this is not stated in the terms of the contract itself.	Witnesses to provide name and address as well as a signature. Ensure proper and binding execution is achieved by Parties that are not incorporated under the *Companies Act* such as housing associations, partnerships and foreign companies.

CONDITIONS				
Section 1: Definitions and Interpretation				
1.1 Definitions	1.3		*Generally:* Definitions that are not amended, or where the change does not have material effect, are not noted here. Several definitions that had been found in the text of WCD 1998 are now located, or at least referenced, in this part of DB 2005. The definitions also serve as something approaching an index, if the term is defined in the text of the contract this is noted and the clause reference is given.	
Agreement	Articles or Articles of Agreement	Slight amendment: the clause repeats the text of WCD 1998 but makes use of new defined terms and makes slight changes, but the effect is unchanged.	The new defined term incorporates a reference to the new Contract Particulars section of the contract.	
Business Day	N/A	**Major change:** the amendment changes a fundamental process or procedure of the contract compared to WCD 1998.	The definition 'Business Day' is new, however the reckoning of periods of days (clause 1.5) and the definition of 'Public Holiday' are unchanged. This is relevant because many of the time-periods stated in the contract make reference to 'days' not 'Business Days'; consequently users should note which units of time they are dealing with as there are usually consequences for failing to observe the correct time-period.	
Conditions	Conditions	**New definition:** this clause provides a new definition of a term or of a process.		
Contract Documents	N/A	**New definition:** this clause provides a new definition of a term or of a process.	This new clause identifies the documents that are 'Contract Documents'.	

DB 2005 clause and reference title	WCD 1998 clause and reference title	Summary of change	Consequence of change and comments	New action required
Contract Particulars	Appendix 1	**New definition:** this clause provides a new definition of a term or of a process.	This is a new definition describing the Contract Particulars and also noting that they are completed by the Parties.	
Contractor's Design Documents	N/A	**New definition:** this clause provides a new definition of a term or of a process.	This clause provides a broad definition of the items that relate to the design of the Works.	
Contractor's Persons	N/A	**Significant amendment:** this is a new clause that accommodates changes introduced in other clauses.	This useful catch-all definition describes those persons the Contractor is responsible for and makes various other clauses more readily understandable.	
Employer's Persons	N/A	**Significant amendment:** This is a new clause that accommodates changes introduced in other clauses.	This useful catch-all definition describes those persons the Employer is responsible for and makes various other clauses more readily understandable.	
Finance Agreement Funder Funder Rights Funder Rights Particulars	N/A	**New definition:** this clause provides a new definition of a term or of a process.	These new definitions support the new regime for providing rights to third parties.	
Gross Valuation	N/A	**New definition:** this clause provides a new definition of a term or of a process.		
Health and Safety Plan	Health and Safety Plan	Clause redrafted: the clause has been significantly reworded but does not give rise to new obligations.		
Insolvent	N/A	**New definition:** this clause provides a new definition of a term or of a process.	Whilst insolvency was an event of great significance under WCD 1998, it is now a defined term as set out in detail in the termination clause.	
Interest Rate	30.6.4, 30.7	**New definition:** this clause provides a new definition of a term or of a process.	The rate of interest was found at various points in WCD 1998 as and when it arose	

			in the text; whereas in DB 2005 'Interest Rate' is a defined term. DB 2005 fixes the Interest Rate at 5% above the Bank of England Base Rate (as such it is the same as those rates stated at various points in WCD 1998).	
Notice of Non-Completion	Notice pursuant to clause 24.1	**New definition:** this clause provides a new definition of a term or of a process.	DB 2005 gives a name to the notice that the Employer must issue in order to activate its right to liquidated damages. Whist this is practical and helpful, this in itself does not change the operation of the clause.	
P&T Rights P&T Particulars Purchaser	N/A	**New definition:** this clause provides a new definition of a term or of a process.	These new definitions support the mechanism for providing rights to Funders, Purchasers and Tenants.	
Practical Completion Statement	16.1	**New definition:** this clause provides a new definition of a term or of a process.	This is another instance of DB 2005 giving a name to a certificate that was required under WCD 1998.	
Pre-agreed Adjustment	12.4.2	**New definition:** this clause provides a new definition of a term or of a process.	This definition gives a name to a process which was also found in WCD 1998 that aids the Parties when the works otherwise required under Provisional Sums need to be changed.	
Provisional Sum	N/A	**New definition:** this clause provides a new definition of a term or of a process.	This definition of 'Provisional Sum' seems to be particularly wide in its scope.	
Rectification Period	Defects Liability Period	**New definition:** this clause provides a new definition of a term or of a process.	This is the new name for what was, under WCD 1998, called the Defects Liability Period.	
Retention Percentage	Retention Percentage	**Major change:** the rights of the Parties have been changed by this amendment compared to WCD 1998.	Users familiar with WCD 1998 should note that the Retention Percentage is 3% for DB 2005. The WCD 1998 provided for Retention at a rate of 5%.	Adjust standard processes to retain 3%, rather than 5%.

DB 2005 clause and reference title	WCD 1998 clause and reference title	Summary of change	Consequence of change and comments	New action required
Sectional Completion Statement	Modifications for Sectional Completion to clause 16.1	**New definition:** this clause provides a new definition of a term or of a process.	This is another instance of DB 2005 giving a name to a certificate that was required under WCD 1998. The Employer is required to recognise Practical Completion of a Section, to issue a certificate for each Section and then to issue a certificate for the whole of the Works when they are complete.	
Statutory Requirements	6.1.1.2 Statutory Requirements	Clause redrafted: the clause has been significantly reworded, but does not give rise to new obligations.		
Works	Works	Clause redrafted: the clause has been significantly reworded, but does not give rise to new obligations.		
Interpretation				
1.2 Reference to Clauses etc.	1.1	Clause redrafted: the clause has been significantly reworded, but does not give rise to new obligations.	The references have been updated for the new nomenclature of the DB 2005, but the effect is unchanged.	
1.3 Articles etc. to be read as a whole	1.2	Clause redrafted: the clause has been significantly reworded and gives additional clarity, but does not give rise to new obligations.	The clause provides a hierarchy for the elements of the contract. The Agreement and the Conditions are placed at the top of the hierarchy. The Employer's Requirements, Contractor's Proposals and/or Contract Sum Analysis are subordinate to the Agreement and Conditions and their contents will not override them.	
1.4 Headings, reference to persons, legislation etc.	N/A	**Significant change:** this is a new clause that provides a new definition of a term or of a process, but does not cause a change to the rights or obligations of the Parties.	This clause might be termed a collection of legal 'boiler plate', that is, text that ensures that words are not taken out of context – it does not of itself create any new obligations but is of value in supporting the operation of the contract.	

1.5 Reckoning periods of days	1.6, 1.7	No change: the clause repeats the text of WCD 1998.		
1.6 Contracts (Rights of Third Parties) Act 1999	1.9	**Significant change:** the clause repeats most of the text of the WCD 1998, but it is also amended in a way that does affect the operation of the clause.	Save to the extent that the contract specifically grants rights to third parties, the provisions of the *Contracts (Rights of Third Parties) Act* 1999 ('the Third Party Rights Act') are to be excluded, and to this extent the clause repeats WCD 1998. Notwithstanding this, DB 2005 gives the Parties the opportunity to grant rights to Funders, Purchasers and Tenants and this may (depending on the options selected) give rise to a need to consider the Third Party Rights Act in detail, as the act may be the mechanism by which those rights are secured.	
1.7 Giving or service of notices and other documents	1.5	**Significant change:** the clause repeats most of the text of the WCD 1998, but it is also amended in a way that does affect the operation of the clause.	Clause 1.7.1 repeats the text of WCD 1998, however clause 1.7.2 adds a useful back-up position for service of documents in the event that the Parties fail to agree a defined date for service. The effect of the clause is that it ensures that the Parties should always be in a position to ensure receipt of a valid notice for the purpose of the contract.	
1.8 Electronic Communications	1.8	**Major change:** the amendment changes a fundamental process or procedure of the contract compared to WCD 1998.	The supplemental conditions for the Electronic Data Exchange have been deleted. DB 2005 provides that the Parties may agree to exchange certain types of document electronically and provides for the Parties to list such items in the Contract Particulars. If the Parties do not provide such a list, then the contract default position is that communications must be in writing and the consequence would be that electronic documents would have no formal status under the contract.	Decide whether or not to consent to valid notice or communication by email, etc. If this is selected, then the specific items of correspondence and the form of communication must be set down in this section of the Contract Particulars.

JCT 2005 Design and Build Contract (DB 2005)

DB 2005 clause and reference title	WCD 1998 clause and reference title	Summary of change	Consequence of change and comments	New action required
			The JCT has not suggested which of the communications required by the contract might sensibly be transferred electronically. At one end of the sliding scale would be a notice of termination, which it is suggested would not be appropriate for electronic notice; and at the other end, the lower-level enquiries and requests for information where it would be practical. Difficulty would arise with the matters in the middle; for example, perhaps it is appropriate for withholding notices to be sent by email but inappropriate for instructions to be sent by email.	
1.9 Effect of Final Account and Final Statement	30.8.1	Slight amendment: the clause repeats the text of WCD 1998 but makes use of new defined terms and makes slight changes, but the effect is unchanged.	Clause 1.9.1 together with its subsections is (saving the use of certain new defined terms) identical to 30.8.1 of WCD 1998. Both clauses 1.9.2 and 1.9.3 have been slightly redrafted, but are fundamentally unchanged from clause 30.8.2 and 30.8.3 of WCD 1998.	
1.10 Effect of payments other than payment of Final Statement	30.9	Slight amendment: the clause repeats the text of WCD 1998 with slight changes that improve the drafting or syntax, but the effect is unchanged.		
1.11 Applicable law	1.7	Clause redrafted: the clause has been significantly reworded and gives additional clarity, but does not give rise to new obligations.		

Section 2: Carrying out the Works				
Contractor's Obligations				
2.1 General obligations	2.1, 6.1.1.1, 6.1.1.2, 8.1.3	Slight amendment: the clause repeats the text of WCD 1998 with slight changes that improve the drafting or syntax, but the effect is unchanged.	Clauses 2.1.1 and 2.1.4 repeat the text of clause 2.1 of WCD 1998 with slight changes to reflect the use of new definitions. Clause 2.1.1 also contains the obligation to comply with the Health and Safety Plan that was located in 8.1.3 of WCD 1998. Clauses 2.1.2 and 2.1.3 are restatements of the obligations contained in 6.1.1.1 and the latter part of 6.1.1.2 – the obligations contained are themselves unchanged.	
2.2 Materials, goods and workmanship	8.1.1, 8.1.2, 8.2, 8.6	**Major change:** the amendment changes a fundamental process or procedure of the contract compared to WCD 1998.	If the Parties describe the standards of workmanship required, then the clause is the same as clauses 8.1 and 8.2 of WCD 1998. That is to say, there are no changes to the standard of workmanship required save in the circumstances where the Parties do not describe the standard, in which case the position has changed substantially. This situation would apply where the Employer fails to state a requirement in his Employer's Requirements and the Contractor then fails to state a standard in the Contractor's Proposals. WCD 1998 provided that in such a situation the 'workmanship shall be of a standard appropriate to the Works'. DB 2005 provides that the Works shall be 'as described in the Contractor's Proposals or documents referred to in clause 2.8'. That clause requires the Contractor to make use of the new Contractor's Design Submission Procedure. The Contractor's	

137

DB 2005 clause and reference title	WCD 1998 clause and reference title	Summary of change	Consequence of change and comments	New action required
			Design Submission is found in Schedule 1 and is analysed below, however for the present purposes it should be noted that the procedure concludes with the establishment of a clear course of action to be taken by the Contractor and does not interfere with the Contractor's responsibility and liability for design. Clause 2.2.3 repeats the obligation found in clause 8.6 of WCD 1998 that the Employer may state that a sample of work will be required of the Contractor and that the Works are to be required to comply with that standard. Note also the update of the terminology used in 2.2.4: the Contractor is required to provide 'reasonable proof' of compliance rather than the 'vouchers' referred to in clause 8.2 of WCD 1998.	
Possession				
2.3 Date of possession – progress	23.1, 23.3.1	Slight amendment: the clause repeats the text of WCD 1998 with slight changes that improve the drafting or syntax, but the effect is unchanged.	The clause is unchanged in substance – in terms of layout, the provisions for sectional completion are included within the main body of the text should they be needed, rather than having them set out in an appendix at the end of the contract.	
2.4 Deferment of possession	23.1.2	Clause redrafted: the clause has been significantly reworded, but does not give rise to new obligations.		
2.5 Early use by Employer	23.3.2, 23.3.3	Slight amendment: the clause repeats the text of WCD 1998 with slight changes that improve the drafting or syntax, but the effect is unchanged.		

2.6 Work not forming part of the Contract	29.1, 29.2	Clause redrafted: the clause has been significantly reworded but does not give rise to new obligations.		
Supply of Documents, Setting Out etc.				
2.7 Contract Documents	5.1, 5.2, 5.4, 5.6	Slight amendment: the clause repeats the text of WCD 1998 but makes use of new defined terms and makes slight changes, but the effect is unchanged.		
2.8 Construction Information	5.3	**Major change:** the amendment changes a fundamental process or procedure of the contract compared to WCD 1998.	The obligation to provide drawings and Design Documents takes effect as and when necessary, i.e. whenever new drawings are required to carry out the Works. The clause requires the Contractor to comply with the new Contractor's Design Submission Procedure. The Contractor must have complied with the Contractor's Design Submission Procedure in order to use any documents that he produces to carry out the Works, and such compliance can only be gained at the end of the Contractor's Design Submission Process. This is the case irrespective of the fact that the obligation to provide the Works may have also arisen.	Contractor's Design Submission Procedure to be applied each time new documents are generated for use on site.
2.9 Site boundaries	7	No change: the clause repeats the text of WCD 1998.		
Discrepancies and Divergences				
2.10 Divergence in Employer's Requirements and definition of site boundary	2.3.1, 2.3.2	Slight amendment: the clause repeats the text of WCD 1998 with slight changes that improve the drafting or syntax, but the effect is unchanged.		

139

DB 2005 clause and reference title	WCD 1998 clause and reference title	Summary of change	Consequence of change and comments	New action required
2.11 Preparation of Employer's Requirements		**Major change:** the rights of the Parties have been changed compared to WCD 1998.	The clause addresses the point identified in *Cooperative Insurance Society Ltd v Henry Boot (Scotland) Ltd* (84 Con LR 164) where Judge Richard Seymour QC said: 'Someone who undertakes … an obligation to complete a design by someone else agrees that the result, however much of the design work was done before the process of completion commenced, will have been prepared with reasonable skill and care.' This clause of DB 2005 seeks to establish that such an obligation will not be assumed by the Contractor.	Clause 2.11 is a specific amendment in respect of the Contractor's design liability made by the JCT to give effect to the original intention and purpose of the design liability clause.
2.12 Employer's Requirements – inadequacy	2.4.1	Clause redrafted: the clause has been significantly reworded and gives additional clarity, but does not give rise to new obligations.	The redrafted clause is much shorter but has the same effect as under WCD 1998.	If the Employer's Requirements are deficient and the Contractors Proposals have not addressed the inadequacy, then, the contract Change process applies.
2.13 Notification of discrepancies etc.	2.4.1, 2.4.3	Clause redrafted: the clause has been significantly reworded and gives additional clarity, but does not give rise to new obligations.	The requirement has been redrafted – the Contractor must now provide 'appropriate details' whereas WCD 1998 required that the notice 'specified the discrepancy'.	It is not anticipated that the consequence of this change will be significant.
2.14 Discrepancies in documents	2.4.1, 2.4.2	Slight amendment: the clause repeats the text of WCD 1998 with slight changes that improve the drafting or syntax, but the effect is unchanged.		
2.15 Divergences from Statutory Requirements	6.1.2, 6.3.1, 6.3.2, 6.3.3	Slight amendment: the clause repeats the text of WCD 1998 but makes use of new defined terms and makes slight changes but the effect is unchanged.	Note that 2.15.2.1 is a simplified redraft of clause 6.3.1 of WCD 1998.	
2.16 Emergency compliance with Statutory Requirements	6.1.3.1	Slight amendment: the clause repeats the text of WCD 1998 with slight changes that improve the drafting or syntax, but the effect is unchanged.		
2.17 Design Work – liabilities and limitation	2.5.1, 2.5.2, 2.5.3	Slight amendment: the clause repeats the text of WCD 1998 with slight changes that improve the drafting or syntax, but the effect is unchanged.		

Fees, Royalties and Patent Rights			
2.18 Fees or charges legally demandable	6.2	Slight amendment: the clause repeats the text of WCD 1998 with slight changes that improve the drafting or syntax, but the effect is unchanged.	
2.19 Royalties and patent rights – Contractor's indemnity	9.1	Slight amendment: the clause repeats the text of WCD 1998 with slight changes that improve the drafting or syntax, but the effect is unchanged.	
2.20 Patent Rights – Instructions	9.2	Slight amendment: the clause repeats the text of WCD 1998 with slight changes that improve the drafting or syntax, but the effect is unchanged.	
Unfixed Materials and Goods – property, risk etc.			
2.21 Materials and goods – on site	15.1	Slight amendment: the clause repeats the text of WCD 1998 with slight changes that improve the drafting or syntax, but the effect is unchanged.	Note that although the title to the clause has changed, the content is exactly the same and of the same effect.
2.22 Materials and goods – off site	15.3	Slight amendment: the clause repeats the text of WCD 1998 with slight changes that improve the drafting or syntax, but the effect is unchanged.	
Adjustment of Completion Date			
2.23 Related definition and interpretation	15.1	**New definitions:** this clause provides a new definition of a term or of a process.	The definitions themselves are uncontroversial.
2.24 Notice by Contractor of delay to progress	25.2.2	Slight amendment: the clause repeats the text of WCD 1998 with slight changes that improve the drafting or syntax, but the effect is unchanged.	

141

DB 2005 clause and reference title	WCD 1998 clause and reference title	Summary of change	Consequence of change and comments	New action required
2.25 Fixing Completion Date 2.25.1	25.3.1.1, 25.3.1.2	**Significant change:** the clause repeats most of the text of the WCD 1998, but it is also amended in a way that does affect the operation of the clause.	The period by which the Completion Date is adjusted remains an estimate of what is 'fair and reasonable' at that time of the assessment. The Employer's duty to adjust the date for completion was previously stated to be conditional upon the Contractor having provided notice, particulars and an estimate of the delay (clause 25.2.2 of WCD 1998); this same provision is repeated in clause 2.24.1.2 of DB 2005. It appears that under DB 2005 an estimate is not required before the obligation to act becomes effective upon the Employer. This is material, as the definition of 'notice' for this clause does refer to the provision of an estimate but this is couched in terms of its practicability (clause 2.24.2) and thus the obligation on the Employer to act may arise before the obligation on the Contractor to provide the estimate has arisen. In consequence, the Employer may well reach a point where he is compelled to act and fix a Completion Date rather than issue innumerable requests for estimates and further information.	
2.25.2	25.3.1	**Significant change:** the clause repeats most of the text of the WCD 1998, but it is also amended in a way that does affect the operation of the clause.	The time-period given to the Employer to state his decision remains 12 weeks. DB 2005 states a requirement for the Employer to 'endeavour' to give the decision in less than 12 weeks when the project is less than 12 weeks from completion. This appears less absolute than WCD 1998 where it was stated that the date was to be fixed 'not later than the Completion Date'. In	

			consequence, the Employer must make the decision but is now given the time to make it despite the approach of the Completion Date. In addition, the obligation is to pass this information on, irrespective of whether or not there is an award of an extension of time. The 2005 form recognises that the Employer makes a decision whether or not to adjust the Completion Date. This change of terminology to adopt the word 'decision' is perhaps a slightly more personal term indicating the need for action and suggests a proactive role.	
2.25.3	25.3.1.3, 25.3.1.4	**Major change:** the amendment changes a fundamental process or procedure of the contract compared to WCD 1998.	Under WCD 1998 there was a tendency for notices granting revisions to the Completion Date to be lacking in detail – frequently due to practical difficulties bearing upon the Employer's Agent. The ability to provide such imprecise revisions to the Completion Date has been reduced by the DB 2005 wording of the clause. The new clauses 2.25.3.1 and 2.25.3.2 provide that the Employer shall state the adjustment to the time for completion attributed to each Relevant Event or Relevant Omission. It would not appear to be satisfactory only to list the Relevant Events that have been taken into account. This suggests period of time (as a number of days or weeks) to be stated against each Relevant Event. It is suggested that if a matter is a Relevant Event but no time is due against it, that the Employer must state that a period of zero weeks is due. In order to provide a complete response, it is suggested that it would be necessary for the Employer to state that the matter the Contractor gave notice of was not a Relevant Matter.	The Employer is to provide a response to each notice for an adjustment to the Completion Date from the Contractor. The Employer's response is to state the period of time he has awarded against each Relevant Event.

JCT 2005 Design and Build Contract (DB 2005)

DB 2005 clause and reference title	WCD 1998 clause and reference title	Summary of change	Consequence of change and comments	New action required
			The Employer therefore has to provide a detailed report of his decisions stating: ● which of the matters notified are Relevant Events; ● which events are not Relevant Events; ● what period of time is due against each Relevant Event; ● if it the Employer has assessed that the Relevant Event has not caused delay that the Completion Date will not be Adjusted. The most appropriate way of reporting this may be as a tabulated analysis given that, once granted, a revision to the Completion Date cannot be taken back. Further reductions or additions to the Completion Date would be dependant upon reductions or expansions in the Works. This changes the process for adjusting completion and elevates the procedural standard to an exacting level akin to that required in the instance of a withholding notice. However, the view has been expressed that Contract Administrators presently discharging their duties under the 1998 form of contract ought already to be considering their decisions in such a level of detail – but also that they ought to be stating that consideration in the manner set out in the 2005 version of the contract.	
2.25.4	25.3.2	Clause redrafted: the clause has been significantly reworded, but does not give rise to new obligations.		
2.25.5	25.3.3	Clause redrafted: the clause has been significantly reworded but does not give rise to new obligations.		

2.25.6	25.3.4, 25.3.5	**Significant change:** the clause repeats most of the text of the WCD 1998 but it is also amended in a way that does effect the operation of the clause.	Note that subclause 2.25.6.4 adds a new provision to support a time-period fixed as part of a Pre-Agreed Adjustment.	
2.26 Relevant Events	25.4	**Major change:** the rights of the Parties have been changed by this amendment compared to WCD 1998.	The list of Relevant Events for DB 2005 is shorter than that of WCD 1998. Most of the changes are a result of more economic drafting, but there are differences of substance. Certain Relevant Events have been combined or stated in express terms. For example, the first of the Relevant Events states that an instruction for a Change shall be a Relevant Event (at 2.26.1). This appears to be something of a catch-all clause, but actually restates the provisions of clause 25.4.5.1 of WCD 1998. A small number of WCD 1998 Relevant Events have been deleted. These are 25.4.10.2 and 25.4.10.3 and concern a Contractor's inability to provide labour or materials and might, perhaps, be categorised as hangovers from the economic climate of the 1970s and 1980s. There has been a degree of consolidation, mainly the result of the Relevant Event in respect of the act or omission by the Employer or the Employer's Persons (clause 2.26.5) – this has consolidated five of the WCD 1998 Relevant Events these were: late instructions (etc.) (25.4.6); impediment by non-contract works (25.4.8.1); Employer's failure to provide (25.4.8.2), Employer's failure to give ingress/egress (25.4.12); Employer's failure to comply with CDM Regulations (25.4.16).	

DB 2005 clause and reference title	WCD 1998 clause and reference title	Summary of change	Consequence of change and comments	New action required
Practical Completion, Lateness and Liquidated Damages				
2.27 Practical completion	16.1	Clause redrafted: the clause has been significantly reworded and gives additional clarity, but does not give rise to new obligations.	The redrafted clause makes provision for the selection of the sectional completion option. The clause has also given a name to the statement that is issued and states that such a 'Practical Completion Statement' is to be issued 'forthwith' – both changes are a slight improvement on WCD 1998, which stated that the issue of such a certificate 'shall not be unreasonably delayed'.	
2.28 Non-Completion Notice	24.1	Slight amendment: the clause repeats the text of WCD 1998 with slight changes that improve the drafting or syntax, but the effect is unchanged.	DB 2005 gives a name to the certificate that is issued in the instance of the Contractor's failure to achieve Practical Completion before the Completion Date. Note also that the clause has been moved from its location under WCD 1998 and its new placement with other 'time' and 'completion' clauses is appropriate.	
2.29 Payment or allowance of liquidated damages	2.24.2	Clause redrafted: the clause has been significantly reworded, but does not give rise to new obligations.		
Partial Possession by Employer				
2.30 Contractor's Consent	17.1	No change: the clause repeats the text of WCD 1998.		

2.31 Practical Completion Date	17.1.1	Slight amendment: the clause repeats the text of WCD 1998 with slight changes that improve the drafting or syntax, but the effect is unchanged.	
2.32 Defects etc. – Relevant Part	17.1.2	Slight amendment: the clause repeats the text of WCD 1998 with slight changes that improve the drafting or syntax, but the effect is unchanged.	
2.33 Insurance – Relevant Part	17.1.3	Slight amendment: the clause repeats the text of WCD 1998 with slight changes that improve the drafting or syntax, but the effect is unchanged.	
2.34 Liquidated damages – Relevant part	17.1.4	Clause redrafted: the clause has been significantly reworded and gives additional clarity but does not give rise to new obligations.	The redrafted clause is far simpler and achieves the same effect – i.e. that the sum of liquidated damages is reduced to take account of the partial possession. The amendment does not get around the fact that there may still be a wrangle over the appropriate deduction, but this is again stated more clearly in the new clause than it was under WCD 1998.	
Defects			
2.35 Schedules of defects and instructions	16.2, 16.3	Clause redrafted: the clause has been significantly reworded but does not give rise to new obligations.	The clause retains the same basic provisions as its predecessor, but has an improved layout. The term 'Defects Liability Period' has been superseded by the term 'Rectification Period'. The redrafting of the clause has moderately improved it – the merging of what was previously presented as two clauses demonstrates that the same process is to be applied in each case, and also makes the contract a little bit more concise.	Update any standard correspondence to accommodate new defined terms and terminology.

DB 2005 clause and reference title	WCD 1998 clause and reference title	Summary of change	Consequence of change and comments	New action required
			In addition to such changes of style, the position on frost damage has changed. There is no longer a specific reference to frost damage, whereas under WCD 1998 the Contractor was liable to meet the cost or remediation of damage caused by frost before Practical Completion.	
2.36 Notice of Completion of Making Good	16.4	Slight amendment: the clause repeats the text of WCD 1998 with slight changes that improve the drafting or syntax, but the effect is unchanged.		
Contractor's Design Documents				
2.37 As-build Drawings	5.5	**Significant change:** the clause has been significantly reworded and gives additional clarity, but does not give rise to new obligations.	DB 2005 states that the as-built drawings are to be provided before practical completion (of the Section of Works) emphasising the point that Practical Completion cannot be achieved until such time as these drawings are provided. WCD 1998 expressed that the documents be provided before commencement of the Defects Liability Period – in effect these provisions are the same, but the emphasis is different. The clause also makes use of the new defined term 'Contractor's Design Documents' – this has the dual benefit of being more concise and more precise than the terms of WCD 1998.	
2.38 Copyright and use	5.6	**Major change:** the amendment changes a fundamental process or procedure of the contract compared to WCD 1998.	The extent of the Employer's right to the use of the design and associated intellectual property is stated and is of particular relevance to any contract where the design is provided by the Contractor.	

			In comparison to WCD 1998, the list of permitted uses of the material is more complete in terms of the activities to which the Employer may apply the Contractor's design. The clause is also better in terms of the legal basis on which these rights are granted – specifically, the provision of a (limited) licence for the use of the material. Note that the limitation of the use of the licence also applies to the Contractor's limitation of liability in relation to any application of the design to different projects by the Employer. It should be noted that the Employer's rights are explicitly stated to be conditional upon the Contractor having been paid.	
Section 3: Control of the Works				
Access and Representatives				
3.1 Access for Employer's Agent	11	Slight amendment: the clause repeats the text of WCD 1998 with slight changes that improve the drafting or syntax, but the effect is unchanged.	Note that the provision in relation to access to sub-contractor locations has been relocated to clause 3.4.	
3.2 Person-in-charge	10	Slight amendment: the clause repeats the text of WCD 1998 with slight changes that improve the drafting or syntax, but the effect is unchanged.		
Sub-Letting				
3.3 Consent to sub-letting	18.2.1	Clause redrafted: the clause has been significantly reworded and gives additional clarity, but does not give rise to new obligations.		

149

DB 2005 clause and reference title	WCD 1998 clause and reference title	Summary of change	Consequence of change and comments	New action required
3.4 Conditions of Sub-letting	18.3, 18.2.1	**Significant change:** the clause repeats most of the text of the WCD 1998, but it is also amended in a way that does affect the operation of the clause.	The clause requiring the Contractor to pay interest to a late-paid sub-contractor has been considerably shortened, but is of the same effect as under WCD 1998. Clause 3.4.2.4 is a new clause that requires the Contractor is to ensure warranties are provided by any sub-contractors they are required from. Clause 3.4.2.5 is intended to ensure that the mechanism for the vesting of Listed Items of property of clause 4.15.2.1 is not adversely affected.	
Employer's Instructions				
3.5 Compliance with instructions	4.1.1	Slight amendment: the clause repeats the text of WCD 1998 with slight changes that improve the drafting or syntax, but the effect is unchanged.		
3.6 Non-compliance with instructions	4.1.2	Clause redrafted: the clause has been significantly reworded and gives additional clarity, but does not give rise to new obligations.	The first change is by way of clarification: the contract now states that the Contractor is liable for 'additional cost' of the work being carried out by the other person retained by the Employer – this qualification was not express in WCD 1998. In addition, the sums that would otherwise have been payable to the Contractor are to be deducted from the Contract Sum, whereas clause 4.1.2 of WCD 1998 stated that the sum was recoverable as a debt or by way of deduction. The result is that the Contractor does not have a right to be paid for such work and is liable for the additional cost.	

3.7 Instructions to be in writing	4.3.	Slight amendment: the clause repeats the text of WCD 1998 with slight changes that improve the drafting or syntax, but the effect is unchanged.	Clause 3.7.4 provides that the Employer may issue a written instruction that tidies up 'instructions' that have not been confirmed in writing. DB 2005 states that the instruction may be given to have retrospective effect, whereas WCD 1998 provided that the instruction would be 'deemed to have taken effect on the date on which it was issued otherwise than in writing'. This clause does not address what is to happen if the Contractor complies with a verbal instruction which is not subsequently confirmed in writing.
3.8 Provision empowering instructions	4.2	Clause redrafted: the clause has been significantly reworded but does not give rise to new obligations.	
3.9 Instructions requiring Changes	12.2.1	Clause redrafted: the clause has been significantly reworded, but does not give rise to new obligations.	Clauses 3.9.1–3.9.3 are the redrafted, but substantially unaltered versions of the obligations found at 12.2.1 of WCD 1998. Clause 3.9.4 repeats clause 12.2.2 of WCD 1998 almost verbatim.
3.10 Postponement of work	23.2	Slight amendment: the clause repeats the text of WCD 1998 with slight changes that improve the drafting or syntax, but the effect is unchanged.	The provision has not been substantially altered, however the WCD 1998 version of the clause was arguably more restrictive in its scope as it said that the power to instruct a postponement applied to 'design and construction activity'. DB 2005 form has wider scope as it applies to the postponement of 'any work be executed under this Contract'.
3.11 Instructions on Provisional Sums	12.3	Slight amendment: the clause repeats the text of WCD 1998 with slight changes that improve the drafting or syntax, but the effect is unchanged.	

DB 2005 clause and reference title	WCD 1998 clause and reference title	Summary of change	Consequence of change and comments	New action required
3.12 Inspection – tests	8.3	Slight amendment: the clause repeats the text of WCD 1998 with slight changes that improve the drafting or syntax, but the effect is unchanged.		
3.13 Work not in accordance with the Contract	8.4	Clause redrafted: the clause has been significantly reworded, but does not give rise to new obligations.		
3.14 Workmanship not in accordance with the Contract	8.5	**Significant change:** the clause repeats most of the text of the WCD 1998, but it is also amended in a way that does affect the operation of the clause.	The clause has been amended to make express reference to the Contractor's obligation to carry out the Works in accordance with the Health and Safety Plan.	
Antiquities				
3.15 Effect of antiquities	34.1	Slight amendment: the clause repeats the text of WCD 1998 with slight changes that improve the drafting or syntax, but the effect is unchanged.		
3.16 Instructions on antiquities	34.2	Slight amendment: the clause repeats the text of WCD 1998 with slight changes that improve the drafting or syntax, but the effect is unchanged.	The definition of the term 'Employer's Persons' has removed the need for the second sentence of clause 34.2, as a person authorised to access the works would fall within this definition.	
3.17 Loss and expense arising	34.3.1	Slight amendment: the clause repeats the text of WCD 1998 with slight changes that improve the drafting or syntax, but the effect is unchanged.		

CDM Regulations

3.18 Undertakings to comply	6A.1, 6A.2, 6A.3 6A.5.1, 6A.5.2	**Major change:** the amendment changes a fundamental process or procedure of the contract compared to WCD 1998.	The CDM compliance regime has been redrafted. The Parties exchange obligations to 'duly comply with the CDM Regulations' at clause 3.18 and the remainder of the clause states how this is to apply depending on the identity of the Planning Supervisor or Principal Contractor. Clause 3.18.1 repeats the obligations stated in clause 6A.1 of WCD 1998; and clause 3.18.2 restates the obligations of clauses 6A.2 and 6A.5.1 of WCD 1998. Clause 3.18.3 restates the obligations of clause 6A.3 of WCD 1998 that apply when the Contractor is the Principal Contractor. The main change is that DB 2005 adds the new express obligation that the Principal Contractor is to develop and then issue the Health and Safety Plan to the Employer before construction begins.	
3.19 Appointment of successors	1.4, 6A.4	Clause redrafted: the clause has been significantly reworded and gives additional clarity, but does not give rise to new obligations.		
Section 4: Payment				
Contract Sum and Adjustments				
4.1 Adjustment only under the Conditions	13	Slight amendment: the clause repeats the text of WCD 1998 with slight changes that improve the drafting or syntax, but the effect is unchanged.		

153

DB 2005 clause and reference title	WCD 1998 clause and reference title	Summary of change	Consequence of change and comments	New action required
4.2 Items included in adjustments	30.5.3	Clause redrafted: the clause has been significantly reworded and gives additional clarity but does not give rise to new obligations.	The clause consolidates several clauses of WCD 1998 into a schedule of the items to be included in adjustments to the Contract Sum. The provisions of the new clause are essentially unchanged. It should be noted that clause 30.5.3.13 of WCD 1998 has been deleted – that clause provided for an amount to be added to the Contract Sum in lieu of ascertainment.	
4.3 Taking adjustments into account	3	Slight amendment: the clause repeats the text of WCD 1998 with slight changes that improve the drafting or syntax, but the effect is unchanged.		
Certificates and Payments				
4.4 VAT	14.2, 14.3	Clause redrafted: the clause has been significantly reworded and gives additional clarity, but does not give rise to new obligations.		
4.5 Construction Industry Scheme (CIS)	30A, 31.2	Clause redrafted: the clause has been significantly reworded and gives additional clarity, but does not give rise to new obligations.		
4.6 Advance Payment	30.1.1, 30.1.2	Clause redrafted: the clause has been significantly reworded and gives additional clarity, but does not give rise to new obligations.		
4.7 Issue of Interim Payment	30.1.1.1	Clause redrafted: the clause has been significantly reworded and gives additional clarity, but does not give rise to new obligations.		

4.8 Amounts Due in Interim Payments	30.1.2	Slight amendment: the clause repeats the text of WCD 1998 but makes use of new defined terms and makes slight changes but the effect is unchanged.	The drafting has been improved by the addition found in the first sentence of the clause that makes the express provision that the deductions from the sums due are aggregated. This does not change the basis of calculating the amount due in comparison to WCD 1998 and appears to have been added to avoid doubt. In addition, the term 'Gross Valuation' has now been defined.	
4.9 Application by Contractor	30.3.1	Slight amendment: the clause repeats the text of WCD 1998 with slight changes that improve the drafting or syntax, but the effect is unchanged.		
4.10 Interim Payments				
4.10.1	30.3.6	Slight amendment: the clause repeats the text of WCD 1998 with slight changes that improve the drafting or syntax, but the effect is unchanged.		
4.10.2	30.4.3	**Significant change:** the clause repeats most of the text of the WCD 1998, but it is also amended in a way that does affect the operation of the clause.	The DB 2005 version of this clause includes an express statement that the Employer withholds and/or deducts monies due to the Contractor against other monies due – the difference between this clause and clause 30.3.6 of WCD 1998 is that the previous form made no express reference to withholding.	
4.10.3	30.3.3	Clause redrafted: the clause has been significantly reworded and gives additional clarity but does not give rise to new obligations.	The text has been slightly redrafted and the provision concerning the statement that the Employer 'shall pay the amount proposed …' is relocated to 4.10.5. Users should note that the payment notice under clause 4.10.3 is a valid method of the Employer advising the Contractor that it will pay less than the sum stated in the Contractor's Application for Interim	

DB 2005 clause and reference title	WCD 1998 clause and reference title	Summary of change	Consequence of change and comments	New action required
			Payment. In this respect it is a valid withholding notice, but only if it provides the requisite details as the sums withheld and the reasons for the withholding. If correctly drafted then such a notice could stand on its own and no further notice would be required.	
4.10.4	30.3.4	Slight amendment: the clause repeats the text of WCD 1998 with slight changes that improve the drafting or syntax, but the effect is unchanged.	Clause 4.10.4 contains the provisions for a withholding notice. The notice required must state the sums together with the grounds for the withholding. The notice described complies with the format in the *Housing Grants, Construction and Regeneration Act* 1996 and has not changed. There is no precondition to this clause – thus a notice can be issued under this clause 4.10.4 even if no notice was served under clause 4.10.3.	
4.10.5	30.3.5	**Major change:** the amendment changes a fundamental process or procedure of the contract compared to WCD 1998.	Clause 4.10.5 provides that the Employer may only pay a sum different to that determined in accordance with clause 4.8 if he issues a certificate under clause 4.10.3, in which case he must pay that sum. The Employer may then only pay a sum different to that in his statement under 4.10.3 if he submits a withholding notice under 4.10.4, in which case he may validly pay the sum stated in the certificate under 4.10.4. The contract places no precondition on the issue of a withholding notice under clause 4.10.4 and thus the Employer may issue such a notice, even if a notice has not been issued under 4.10.3.	

			There is further change of great significance in this clause. This change has relevance where there has been a breakdown in the notice process. Under WCD 1998, if the Employer failed to issue any notice before withholding sums, then clause 30.3.5 provided that the Employer was to pay the amount stated in the Contractor's Application for Interim Payment. Whilst this sum was itself supposed to be based on the WCD 1998 formulation of the amount due, it was liable to be the subject of significant controversy in such circumstances. DB 2005 deals with the same problem in a different manner. If the Employer fails to issue the requisite notices, the contract provides that the Employer is to pay 'the amount due … as determined in accordance with clause 4.8'. Whist this may still be an invitation to an adjudication, the question that may be referred to the Adjudicator in such a circumstance would be less fettered by the circumstance and would go to the question of 'what is the amount due?'.	
4.10.6	30.3.7	Slight amendment: the clause repeats the text of WCD 1998 but makes use of new defined terms and makes slight changes, but the effect is unchanged.		
4.11 Contractor's right of suspension	30.3.8	Slight amendment: the clause repeats the text of WCD 1998 but makes use of new defined terms and makes slight changes, but the effect is unchanged.		

DB 2005 clause and reference title	WCD 1998 clause and reference title	Summary of change	Consequence of change and comments	New action required
4.12 Final Account and Final Statement – submission and payment	30.5.1, 30.6	Slight amendment: the clause repeats the text of WCD 1998 but makes use of new defined terms and makes slight changes, but the effect is unchanged.	This large clause made up of 12 subclauses is fundamentally unaltered in comparison to the WCD 1998 form. Such amendments as are found do not change the calculation of the Final Account, nor the process that may be followed in its absence. Most of the amendments have been introduced to adopt the updated defined terms used in DB 2005.	
Gross Valuation				
4.13 Ascertainment – Alternative A	30.2A.1, 30.2A.4	Slight amendment: the clause repeats the text of WCD 1998 but makes use of new defined terms and makes slight changes, but the effect is unchanged.	Notwithstanding the change in layout and use of some new defined terms, the clause is not changed in comparison to WCD 1998.	
4.14 Ascertainment Alternative B	30.2B	Slight amendment: the clause repeats the text of WCD 1998 but makes use of new defined terms and makes slight changes, but the effect is unchanged.	As with clause 4.13, the function of the clause has not changed in comparison to WCD 1998.	
4.15 Off-site materials and goods	15.2	Slight amendment: the clause repeats the text of WCD 1998, but makes use of new defined terms and makes slight changes, but the effect is unchanged.	As with WCD 1998 a bond in respect of off-site materials is not mandatory under DB 2005. A bond will only be required if the Employer has stated that this is a requirement in the Contract Particulars.	
Retention				
4.16 Rules on treatment of Retention	30.4.2	Slight amendment: the clause repeats the text of WCD 1998 with slight changes that improve the drafting or syntax, but the effect is unchanged.		

4.17 Retention – rules for ascertainment	30.4.1	Clause redrafted: the clause has been significantly reworded and gives additional clarity, but does not give rise to new obligations.	The rate of Retention is that stated in the Contract Particulars. Users should note that the default Retention sum has changed from the 5% Retention of WCD 1998 to 3% for DB 2005.	
Fluctuations				
4.18 Choice of provisions	35.1	Slight amendment: the clause repeats the text of WCD 1998 with slight changes that improve the drafting or syntax, but the effect is unchanged.		
Loss and Expense				
4.19 Matters materially affecting regular progress	26.1	Clause redrafted: the clause has been significantly reworded and gives additional clarity, but does not give rise to new obligations.	Note that the statement that the Contractor may give a quantification of the sums he considers are his loss and/or his expenses has been omitted from DB 2005.	
4.20 Relevant Matters				
4.20.1	26.2	Clause redrafted: the clause has been significantly reworded and gives additional clarity but does not give rise to new obligations.	DB 2005 has a shorter list of Relevant Matters than WCD 1998. Clause 4.20.5 addresses many of the points that the previous forms stated as separate subclauses. In consolidating the liabilities and impediments arising from the involvement of the Employer and the Employer's Persons, the clause sweeps up into a single Relevant Matter what were six, separately stated, Relevant Matters in WCD 1998. The 'impediment, prevention or default' referred to in DB 2005 clause 4.20.2.5 would include restricted site access, non-provision of materials or non-compliance with regulations that WCD 1998 stated were Relevant Matters in separate clauses.	

DB 2005 clause and reference title	WCD 1998 clause and reference title	Summary of change	Consequence of change and comments	New action required
4.21 Amounts ascertained – addition to Contract Sum	26.3	Slight amendment: the clause repeats the text of WCD 1998 with slight changes that improve the drafting or syntax, but the effect is unchanged.		
4.22 Reservation of Contractor's rights and remedies	26.4	No change: the clause repeats the text of WCD 1998.		
Section 5: Changes				
General				
5.1 Definition of Changes	12.1	Slight amendment: the clause repeats the text of WCD 1998 with slight changes, but the effect is unchanged.		
5.2 Valuation of Changes and Provisional Sum work	12.4.1, 12.4.2	**Major change:** the amendment changes a fundamental process or procedure of the contract compared to WCD 1998.	In comparison to WCD 1998, DB 2005 has a greatly simplified mechanism for valuing Changes and Provisional Sum work. The clause directs the Parties to reach an agreement over the Valuation of the work as the principal method of setting a price. There is no elaborate mechanism for a 'Contractor's Price Statement' or similar process as was stated in WCD 1998; in consequence, the Parties are free to manage this themselves. If agreement is not possible (and agreement cannot be compelled), the clause still operates to provide a definite outcome as in the absence of agreement there is to be a Valuation in accordance with the Valuation Rules.	

5.3 Giving effect to Valuations, Agreements etc.	12.6	Slight amendment: the clause repeats the text of WCD 1998 with slight changes that improve the drafting or syntax, but the effect is unchanged.	
The Valuation Rules			
5.4 Measurable Work	12.5.1, 12.5.2, 12.5.3	**Significant change:** the clause repeats most of the text of the WCD 1998, but it is also amended in a way that does affect the operation of the clause.	An extra subclause has been added to DB 2005 in comparison to WCD 1998 as clause 5.4.1 and provides that: 'Allowance shall be made in such Valuations for the addition or omission of the relevant design work.' This addition should ensure that both cost and savings associated with changes requiring more or less design work will be reflected in Valuations – this is of relevance as many Contractors make use of the services of a sub-contracted (or usually novated) designer.
5.5 Daywork	12.5.4	Slight amendment: the clause repeats the text of WCD 1998 with very slight changes, but the effect is unchanged.	
5.6 Change of conditions for other work	12.5.5	Clause redrafted: the clause has been significantly reworded and gives additional clarity but does not give rise to new obligations.	
5.7 Additional provisions	12.5.6	Clause redrafted: the clause has been significantly reworded and gives additional clarity, but does not give rise to new obligations.	

DB 2005 clause and reference title	WCD 1998 clause and reference title	Summary of change	Consequence of change and comments	New action required
Section 6: Injury, Damage and Insurance				
Injury to Persons and Property				
6.1 Liability of Contractor – personal injury or death	Clause 20.1	Clause redrafted: the clause has been significantly reworded and gives additional clarity, but does not give rise to new obligations.	The clause has been redrafted – it makes use of the new definition 'Employer's Persons'. More significantly, the DB 2005 version of this clause removes the statement that the liability indemnified is 'under any statute or common law' – in so doing, it removes any restriction that might conceivably have followed from such a statement.	
6.2 Liability of Contractor – Injury or damage to property	20.2	Slight amendment: the clause repeats the text of WCD 1998 but makes use of new defined terms and makes slight changes, but the effect is unchanged.		
6.3 Injury or damage to property – Works and Site Materials excluded	20.3	Clause redrafted: the clause has been significantly reworded and gives additional clarity but does not give rise to new obligations.		
Insurance against Personal Injury and Property Damage				
6.4 Contractor's insurance of his liability	21.1.1.1, 21.1.1.2	**Significant change:** the clause repeats most of the text of the WCD 1998, but it is also amended in a way that does affect the operation of the clause.	The parallel clause in WCD 1998 made specific reference to persons with contracts of service and apprenticeships, whereas this version refers to employees only.	Confirm that apprenticed staff are covered and consider whether self-employed individuals are covered.

			In response to this, Parties should ensure that their apprentices are adequately insured as employees and might wish to consider the status of self-employed staff.	
6.5 Contractor's insurance of liability of Employer	21.2.1, 21.1.2, 21.2.3	Slight amendment: the text of WCD 1998 is repeated but new defined terms are used, slight changes are made and the clause is amended to accommodate changes introduced in other clauses.		
6.6 Excepted Risks	21.3	No change: the clause repeats the text of WCD 1998.		
Insurance of the Works			It is essential that the Contract Particulars relating to insurance are completed, as not all of the Particulars have a default position.	
6.7 Insurance Options	22.1	No change: the clause repeats the text of WCD 1998.	The selection of the insurance regime for the project is made in the Contract Particulars and the Options are relocated to Schedule 3.	
6.8 Related Definitions	22.2	Slight amendment: the clause repeats the text of WCD 1998 with slight changes that improve the drafting or syntax, but the effect is unchanged.	There are a small number of changes to the terms, but the effect is minimal. Some widen the scope – for example, the mechanism or method for the escape of water is no longer prescribed in the definition of Specified Peril (which is sensible). Others reflect an update in the language used – for example, 'tempest' is no longer a Specified Peril, presumably on the grounds that 'storm' is now thought to adequately describe the same Specified Peril. Others are points of clarification – for example, the definition of Joint Names Policy now provides that the Parties are 'composite insured'.	

DB 2005 clause and reference title	WCD 1998 clause and reference title	Summary of change	Consequence of change and comments	New action required
6.9 Sub-contractors – Specified Perils cover under Joint Names All Risks Policies	22.3	**Significant change:** the clause repeats most of the text of the WCD 1998, but it is also amended in a way that does affect the operation of the clause.	This clause ensures that sub-contractors are adequately addressed when the insurance arrangements are put in place – irrespective of whether it is the Employer or the Contractor who is charged with providing the insurance. In essence, sub-contractors will either gain the benefit of the insurance cover as a joint named insured or they are granted a waiver in respect of any subrogation rights. The new clause has a slightly wider scope that requires either recognition of insurance cover or a waiver of subrogation in respect of an additional item – 'work executed'; this was not provided in WCD 1998. The provision has been redrafted to accommodate the fact that sectional completion amendments are no longer found in the Annex.	
6.10 Terrorism Cover – non-availability – Employer's options	22A.5.1, 22B.4.1–3, 22C.5.1	**Significant change:** the clause imports the text from one of the JCT supplements that previously had to be specifically adopted by the Parties.	The insuring Party has to advise the other Party that Terrorism Cover is about to lapse. At this point the Employer has an option to carry on without the insurance or terminate the contract. WCD 1998 stated the process to be followed in the event of non-availability of Terrorism Cover three times (once in each of the insurance options); DB 2005 does it once, but in addition to consolidating the text makes certain changes. The most significant change in the clause is that the mechanism of loss by terrorism has been de-specified – that is to say, the loss is no longer limited to that caused by 'fire or explosion caused by terrorism'.	

			DB 2005 defines it in terms of 'physical loss or damage caused by terrorism' ('terrorism' is not itself a defined term). The amendment greatly widens the range of events that ought to be covered by the insurance. Clause 6.10.4.3 is a new provision against the imposition of a liability to the effect that the Employer has to rebuild the entirety of the damaged structure in order to facilitate the Contractor's obligation to continue with the Works.	
Professional Indemnity Insurance				
6.11 Obligation to insure	N/A	**Major change:** the amendment changes a fundamental process or procedure of the contract compared to WCD 1998.	This is a new clause that provides for a term that will provide an increased measure of security for the Employer. The clause requires the Contractor to insure against failure to discharge its duties to a professional standard. In the context of a design and build contract this is most obviously in respect of design liability. Depending on the terms of the contract such insurance may equally be required in respect of the design of mechanical and electrical works, systems or indeed the installation of specialist plant.	Contractor to obtain and maintain professional indemnity insurance.
6.12 Increased cost and non-availability	N/A	**Major change:** the amendment changes a fundamental process or procedure of the contract compared to WCD 1998.	If it becomes un-commercial for the Contractor to provide professional indemnity insurance, the Parties are to meet to discuss protecting their interests. Although the terms of this clause are vague and would be difficult to enforce (and as they provide little certainty), they are phrased in a widely used form. The alternative to such an arrangement would be to require the insurance for a period of time irrespective of the premium due, or whether the risk justified the price.	

DB 2005 clause and reference title	WCD 1998 clause and reference title	Summary of change	Consequence of change and comments	New action required
Joint Fire Code – Compliance				
6.13 Application of clauses	22FC	No change: the clause repeats the text of WCD 1998.		
6.14 Compliance with Joint Fire Code	22FC.2.1, 22FC.2.2	Clause redrafted: the clause has been significantly reworded and gives additional clarity, but does not give rise to new obligations.		
6.15 Breach of Joint Fire Code – Remedial Measures	22FC.3.1, 22FC.3.2	**Significant change:** the clause repeats most of the text of the WCD 1998, but it is also amended in a way that does affect the operation of the clause.	The recovery of sums from the Contractor in the event of the Contractor's failure to carry out remedial measures has been tightened. Clause 6.15 states that 'all additional costs incurred by the Employer' may be recovered, whereas WCD 1998 provided a less stringent requirement that 'all costs incurred in connection with such employment' could be recovered.	
6.16 Joint Fire Code – amendments/revisions	22FC.5, Appendix 1	Clause redrafted: the clause has been significantly reworded, but does not give rise to new obligations.		
Section 7: Assignment, third party rights and collateral warranties				
Assignment				
7.1 General	18.1	**Significant change:** the clause repeats most of the text of the WCD 1998, but it is also amended in a way that does affect the operation of the clause.	The drafting has been tidied up to make it clear that the provisions of clauses 7.1 and 7.2 are not contradictory.	

7.2 Rights of enforcement	18.1, 18.2	Slight amendment: the text of WCD 1998 is repeated but new defined terms are used, slight changes are made and the clause is amended to accommodate changes introduced in other clauses.		
Clauses 7A to 7F – Preliminary			*Generally:* A second section of Contract Particulars deals specifically with third party rights and collateral warranties – it is essential that this is completed, as the default position is that there are to be no such rights unless there is a specific opt-in. The rights themselves are set out in Schedule 5 of the contract and all of the rights are optional.	Complete the third party rights section of the Contract Particulars.
7.3 References		**Major change:** the amendment changes a fundamental process or procedure of the contract compared to WCD 1998.	This term allows the incorporation of further documents or schedules of rights that may be appended to the contract.	
7.4 Notices		**Major change:** the amendment changes a fundamental process or procedure of the contract compared to WCD 1998.	The requirements for giving proper notice under this clause are stated here. It is suggested that if a JCT collateral warranty were to be amended, then the warranty would not be a 'specified JCT collateral warranty' and the amended form would have to accompany the notice.	
7.5 Execution of collateral warranties		**Major change:** the amendment changes a fundamental process or procedure of the contract compared to WCD 1998.	This clause ensures that the collateral warranties required follow the main contract so that a warranty by deed is not required where the main contract is under hand and vice versa. Most readers will be aware that the limitation period for an action for a breach of contract for a agreement executed under hand is six years and that the period is 12 years for a deed.	

DB 2005 clause and reference title	WCD 1998 clause and reference title	Summary of change	Consequence of change and comments	New action required
Third party rights from Contractor				
7A Rights for Purchasers and Tenants		**Major change:** the amendment changes a fundamental process or procedure of the contract compared to WCD 1998.	The Contract Particulars must state this requirement. The rights vest at the time the Employer gives the notice to the Contractor. The notice must comply with 7.4 by being in writing and given by actual delivery (i.e. delivered by hand rather than mailed), or by special or recorded delivery. Ordinary post or fax does not suffice. The notice must also comply with 7A.1, which requires the identity of the person acquiring the rights and their interest to be stated – that is to say, the nature of his interest in the Works. Clause 7A.2 notes that the Parties are free to continue to exercise their own rights (such as termination, waiver or settlement) under the contract even though Purchasers and Tenants may have rights too. The Parties can agree to amend the rights that will be given, however clause 7A.3 does prevent the Parties from changing the rights of the Purchasers and Tenants once the P&T rights have been vested.	If Purchaser and Tenant rights are required, then this requirement is to be stated in the Contract Particulars before the contract is executed.
7B Rights for a Funder		**Major change:** the amendment changes a fundamental process or procedure of the contract compared to WCD 1998.	The Contract Particulars must state this requirement. The rights vest at the time the Employer gives the notice to the Contractor; once vested the Funder can act in accordance with the rights that he has been given from that point. The notice must comply with 7.4 by being in writing and given by actual delivery	If Funder's rights are required, then this requirement is to be stated in the Contract Particulars before the contract is executed.

		(i.e. delivered by hand rather than mailed), or by special or recorded delivery. Ordinary post or fax does not suffice. The notice must also comply with clause 7B.1 which requires the identity of the Funder acquiring the rights to be stated. The clause permits the Parties to operate the contract without needing to get the consent of the Funder – thus variation and settlement of items in dispute and waiver of terms are not restricted. However (unlike the P&T rights), the rights of the Funder do place certain limitations on the Parties in some respects. Specifically, clause 7B.2 prevents either Party from rescinding the contract.	
Collateral warranties			
7C Contractor's Warranties – Purchasers and Tenants	**Major change:** the amendment changes a fundamental process or procedure of the contract compared to WCD 1998.	The Contract Particulars must state that collateral warranties for Purchasers and Tenants are a requirement for this clause to give rise to an obligation on the Contractor. The obligation to procure the collateral warranties arises at the time the Employer gives the notice to the Contractor. The notice must comply with 7.4 by being in writing and given by actual delivery (i.e. delivered by hand rather than mailed), or by special or recorded delivery. Ordinary post or fax does not suffice. The notice must also enclose the terms of the Warranty if it is different or has amendments to the JCT collateral warranty (7.4). The Contractor must enter into the warranty within 14 days of receiving the notice.	If Contractor's Warranties are required in favour of Purchasers and Tenants rights, then this is to be stated in the Contract Particulars before the contract is executed.

DB 2005 clause and reference title	WCD 1998 clause and reference title	Summary of change	Consequence of change and comments	New action required
7D Contractor's Warranty – Funder		**Major change:** the amendment changes a fundamental process or procedure of the contract compared to WCD 1998.	The Contract Particulars must state that collateral warranties for a Funder are a requirement for this clause to give rise to an obligation. The obligation to procure the Warranties arises at the time the Employer gives the notice to the Contractor. The notice must comply with 7.4 by being in writing and given by actual delivery (i.e. delivered by hand rather than mailed), or by special or recorded delivery. Ordinary post or fax does not suffice. The notice must also enclose the terms of the Warranty if it is different or has amendments to the JCT collateral warranty (7.4). The Contractor must enter into the Warranty within 14 days of receiving the notice.	If Contractor's Warranties are required in favour of the Funder, then this requirement is to be stated in the Contract Particulars before the Contract is executed.
7E Sub-contractor's warranties – Purchasers and Tenants/Funder		**Major change:** the amendment changes a fundamental process or procedure of the contract compared to WCD 1998.	The Contract Particulars must state that collateral warranties are required from consultants or sub-contractors in favour of Funders and/or Purchasers and Tenants for this clause to give rise to an obligation. The obligation to provide the collateral warranty takes effect upon the receipt of a notice that complies with clause 7.4 by being in writing and given by actual delivery (i.e. delivered by hand rather than mailed), or by special or recorded delivery. Ordinary post or fax does not suffice. The notice also has to state which Contractor the warranties are required from, in whose favour they must be and in which form. The notice must also enclose the terms of the collateral	If sub-contractors' warranties are required in favour of the Purchasers and Tenants or Funders, then this requirement is to be stated in the Contract Particulars before the contract is executed.

			warranty if it is different or has amendments to the JCT collateral warranty (7.4). The consultants and sub-contractors may propose amendments and although both the Contractor and the Employer are to approve the amendments they are not permitted to unreasonably delay or withhold consent. The result is that small-scale changes would ultimately have to be accepted. If the Employer states specific requirements in the Contract Particulars, then it is suggested that it would not be unreasonable to withhold consent if the warranty proposed does not satisfy the requirements. The Contractor must achieve the execution of the warranty by its sub-contractor within 21 days of the Employer's notice.	
7F Sub-contractor's warranties – Employer		**Major change:** the amendment changes a fundamental process or procedure of the contract compared to WCD 1998.	The Particulars must state that collateral warranties are required from sub-contractors in favour of the Employer for this clause to take effect. The obligation to provide the warranty takes effect upon the receipt of a notice that complies with clause 7.4 by being in writing and given by actual delivery (i.e. delivered by hand rather than mailed), or by special or recorded delivery. Ordinary post or fax does not suffice. The notice also has to state which Contractor the notice is required from, in whose favour and in which form. The notice must also enclose the terms of the warranty if it is different or has amendments to the JCT collateral warranty (clause 7.4).	If the Employer requires sub-contractors' warranties in his favour, then this requirement is to be stated in the Contract Particulars before the contract is executed.

171

DB 2005 clause and reference title	WCD 1998 clause and reference title	Summary of change	Consequence of change and comments	New action required
			The Contractor must achieve the execution of the warranty by its sub-contractor within 21 days of the Employer's notice. The clause specifically considers that the warranty may be provided in different terms. The Employer's approval of the amendments is required, however he cannot unreasonably delay or withhold the consent. The result is that small-scale changes would ultimately have to be accepted by the Employer. If the Employer has states specific requirements in the Contract Particulars, then it is suggested that it would not be unreasonable to withhold consent if the warranty proposed does not satisfy the requirements.	
Section 8: Termination			*Generally:* The first change noted will be that a contract is 'terminated' rather than 'determined'. In general the clause is more concise, and a good deal of repetition has been avoided. In operation and effect DB 2005 is very similar to WCD 1998. Many of the changes relate to insolvency. Perhaps the most significant change for those used to WCD 1998 is that a notice will always be required to terminate the contract even in the event of insolvency as there is no longer automatic termination for this event. It should be noted that the provision for the payment of sub-contractors directly has been omitted, as has the arrangement for novation or continuation.	

General				
8.1 Meaning of Insolvency	27.3.1	**Major change:** the amendment changes a fundamental process or procedure of the contract compared to WCD 1998.	The clause stating the meaning of 'insolvency' for the purpose of this contract contains many of the elements of the previous version of the contract, however it is a wider definition and includes additional elements. Clause 8.1.1 in particular is of broad reach and would catch many of the preparatory stages of insolvency. Users should note clause 8.1.5, which states that that procedures in other legal jurisdictions also fall within the definition of 'Insolvent' for DB 2005. Thus, procedures analogous to administration or bankruptcy overseas will have direct consequences under the contract. This is of consequence as clause 5.8 provides that certain obligations are suspended upon the instance of insolvency.	
8.2 Notices under section 8	27.2.4, 27.2.3, 27.1	Clause redrafted: the clause has been significantly reworded and gives additional clarity, but does not give rise to new obligations.	Notices of termination cannot be given merely to annoy or vex, nor can they be given unreasonably. It should be noted that this provision applies to both Employer and Contractor, as whichever activates the termination process (whether under Termination by the Employer (clause 8.4) or Termination by the Contractor (clause 8.9)) the ultimate result is 'termination of the Contractor's Employment'. The termination notice takes effect upon the receipt of a notice that complies with clause 8.2. by being in writing and given by actual delivery (i.e. delivered by hand rather than mailed), or by special or recorded delivery. Ordinary post or fax does not suffice. The assumption of delivery on the second Business Day can be rebutted if a Party can demonstrate	

173

DB 2005 clause and reference title	WCD 1998 clause and reference title	Summary of change	Consequence of change and comments	New action required
			that the notice was given sooner or indeed that it was not given at all.	
8.3 Other rights, reinstatement	27.8, 28.5	**Significant change:** the clause repeats most of the text of the WCD 1998, but it is also amended in a way that does affect the operation of the clause.	Clause 8.3.2 adds a note that states that the Parties may agree to a reinstatement incorporating new or additional terms. Whilst reinstatement may make commercial or construction sense, DB 2005 does not create an enforceable obligation even to consider reinstatement.	
Termination by Employer				
8.4 Default by Contractor	27.2.1–27.2.3	Slight amendment: the clause repeats the text of WCD 1998 with slight changes that improve the drafting or syntax, but the effect is unchanged.		
8.5 Insolvency of Contractor	27.3.2, 27.5.1, 27.5.4	**Major change:** the amendment changes a fundamental process or procedure of the contract compared to WCD 1998.	If the Contractor is Insolvent, the Employer can terminate immediately by notice. The onus is therefore upon the Employer to act swiftly and decisively in the instance of the insolvency of the Contractor. This is a major change compared to WCD 1998, where determination occurred immediately and without any action upon the Employer upon the Contractor's insolvency in almost all cases. The Contractor is obliged to tell the Employer if he is approaching insolvency. This clause states the consequences of an insolvency. It is suggested that these provisions would apply irrespective of whether the event of insolvency has been notified to the Employer. By extension,	Notice must be given to the Contractor by the Employer if he wishes to terminate the contract in the event of the Contractor's insolvency.

			this clause equally applies to an insolvency event in a different jurisdiction as defined under clause 8.1.5. Upon Insolvency the elements of the contract between the Parties are effectively suspended – the Contractor's obligation to continue with the Works is suspended, and the Employer's obligation to make further payment ceases to apply and he can secure the site and exclude the Contractor from it.	
8.6 Corruption	27.4	Clause redrafted: the clause has been significantly reworded, but does not give rise to new obligations.	A concise statement that the Contractor's corruption on any project will permit the Employer to terminate the Contract. The clause also makes reference to the *Local Government Act* 1972 to accommodate the obligations as they apply to local authorities.	
8.7 Consequences of termination under clauses 8.4 to 8.6	27.6	**Significant change:** the clause repeats most of the text of the WCD 1998, but it is also amended in a way that does affect the operation of the clause.	The redrafted clause is, on balance, an improvement on WCD 2005. The provision permitting the accounting for, and recovery of, the Employer's loss and expense has been widened to include loss and expense otherwise than as a result of the termination in clause 8.7.2.4.1.	
8.8 Employer's decision not to complete the Works	27.7.1	Clause redrafted: the clause has been significantly reworded and gives additional clarity, but does not give rise to new obligations.		
Termination by Contractor				
8.9 Default by the Employer	28.1	Slight amendment: the clause repeats the text of WCD 1998 with slight changes that improve the drafting or syntax, but the effect is unchanged.		

DB 2005 clause and reference title	WCD 1998 clause and reference title	Summary of change	Consequence of change and comments	New action required
8.10 Insolvency of Employer	28.3	Clause redrafted: the clause has been significantly reworded and gives additional clarity, but does not give rise to new obligations.		
8.11 Termination by either Party	28A.1	Clause redrafted: the clause has been significantly reworded, but does not give rise to new obligations.		
8.12 Consequences of termination under clauses 8.9 to 8.11 etc.	28A.2	**Major change:** the rights of the Parties have been changed by this amendment compared to WCD 1998.	Under WCD 1998 the Employer was obliged to prepare an account and the Contractor was bound to contribute all of the materials within two months of determination. DB 2005 provides that the Contractor is obliged to compile this account himself in instances of default or insolvency of the Employer (clauses 8.9 and 8.10 respectively). The contract also provides for the Employer to elect that the Contractor should compile the report in the event of a suspension that has led to a termination, or termination due to lack of terrorism insurance or termination following damage to the structure. Clause 28A.5 of WCD 1998 provided for a release of 50% of the Retention before the determination of the contract – this clause has not been maintained in DB 2005. Clause 8.12.4 is a new clause that limits the applicability of clause 8.12.3.5, which provides for the recoverability of direct loss and/or damage caused to the Contractor by termination. The amendment allows such recovery only in the event of default by the Employer (or the Employer's Persons) under 8.9 and 8.11.1.3 and instances of the Employer's insolvency.	Contractor to prepare account on termination following Employer's Insolvency or default.

Section 9: Settlement of Disputes				
9.1 Mediation	Footnote (dd)	**Major change:** the amendment changes a fundamental process or procedure of the contract compared to WCD 1998.	This 'recommendation' that the Parties consider mediating their disputes has no binding force and the Parties may not be compelled by this clause to mediate or even give serious consideration to mediating a dispute. However, mediation may, in the particular circumstances of a dispute, be a sensible and cost-effective way to resolve a dispute – the courts certainly approve and are directing many cases in this direction.	
9.2 Adjudication	39A.1 - 39A.8	**Major change:** the amendment changes a fundamental process or procedure of the contract compared to WCD 1998.	The adjudication procedure set out at length in WCD 1998 has been deleted. DB 2005 adopts the Scheme set out in the *Scheme for Construction Contracts (England and Wales) Regulations* 1998 ('the Scheme'). The operation of the Scheme is slightly amended. Firstly, by the provision that the Parties may identify their own Adjudicator or Adjudicator Nominating Body if the necessary amendment is made in the Contract Particulars (this is the same as WCD 1998). Secondly, clause 9.2.2 also departs from the Scheme and provides that the candidate for the role of Adjudicator of a dispute concerning opening up and testing (under clause 3.18.4) must either have specialist knowledge or must appoint a specialist expert to give written advice. This too maintains the provisions of WCD 1998 clause 29A.5.8. Users should familiarise themselves with the Scheme for Construction Contracts.	The Scheme for Construction Contracts should be (re)read before contemplating any adjudication. Terms of the Adjudicator's appointment must be agreed with the Adjudicator.

DB 2005 clause and reference title	WCD 1998 clause and reference title	Summary of change	Consequence of change and comments	New action required
			In addition to the removal of the JCT's adjudication process, it is no longer a condition that the Adjudicator has to accept the JCT Adjudication Agreement that provided terms between the Parties and the Adjudicator – all references to it have been removed. On balance, this is a positive change that has removed a potential difficulty in the Adjudicator selection process. Problems occasionally arose if Adjudicators were not willing to use the JCT Adjudication Agreement.	
Arbitration				
9.3 Conduct of arbitration	39B, 39B.6	**Major change:** the amendment changes a fundamental process or procedure of the contract compared to WCD 1998.	Resolution of disputes by Arbitration is not the default position under DB 2005; consequently, these clauses will not apply unless the Parties make the necessary changes in the Contract Particulars. The contract adopts the latest (2005) version of the Construction Industry Model Arbitration Rules (CIMAR) and these will apply to an arbitration pursuant to Article 8 unless (as the clause suggests) the Parties agree to accept whatever CIMAR are current at the date of the dispute. Note that the Parties to a contract that does not adopt arbitration may still have a dispute arbitrated if they both agree. In such a circumstance the Parties would not be arbitrating 'pursuant to Article 8'; thus, unless they also agreed to accept them, these clauses would not apply and the	

			Parties would also have to agree the appropriate arbitration rules.	
9.4 Notice of reference to arbitration	39B1.1	Clause redrafted: the clause has been significantly reworded, but does not give rise to new obligations.	The method and process for giving notice of arbitration is stated here. The distinction is that WCD 1998 reproduced the relevant part of Rule 2.1 of the CIMAR – DB 2005 does not.	
9.5 Powers of Arbitrator	39B.2	Slight amendment: the clause repeats the text of WCD 1998 with slight changes that improve the drafting or syntax, but the effect is unchanged.		
9.6 Effect of award	39B.3	No change: the clause repeats the text of WCD 1998.		
9.7 Appeal – questions of law	39B.4	No change: the clause repeats the text of WCD 1998.		
9.8 Arbitration Act 1996	39B.5	Slight amendment: the clause repeats the text of WCD 1998 but is amended to incorporate or accommodate changes introduced in other clauses.		
SCHEDULES				
Schedule 1: Contractor's Design Submission Procedure			Unlike WCD 1998, the Design Submission Procedure for DB 2005 is not optional or contained in a supplemental condition – it is mandatory. Thus the process will apply whenever the Contractor is generating Contractor's Design Documents, which are defined as: '… the drawings, details, and specifications of materials goods and workmanship and other related documents prepared by or for the Contractor in relation to the design of the Works'. It will thus apply throughout the design and construction process and will not only apply to any Changes that are instructed.	

179

DB 2005 clause and reference title	WCD 1998 clause and reference title	Summary of change	Consequence of change and comments	New action required
			The process is set up to give certainty, even in the instance of the Employer failing to (or deciding not to) apply the process. Employers must be aware that they cannot play for time or get away with ignoring the Contractor's submissions – if he does, the Contractor will be entitled to get on with the works making use of his design.	
1.	5.3, S2.1	**Major change:** the amendment changes a fundamental process or procedure of the contract compared to WCD 1998.	The paragraph requires two copies of the Contractor's Design Documents to be sent to the Employer. If the Employer has a particular format that it requires the Contractor to adopt, then this must be stated in the Contract Particulars. This is likely where the Employer requires the submission of designs or data using a specific software package or protocol, but may be as practical as the requirement for sequential or coded identification of documents. The requirement that the Contractor's Proposals be submitted 'in sufficient time to allow any comments by the Employer' is subjective and open ended.	
2.	N/A	**Major change:** this is a wholly new clause for the DB 2005.	The Employer must act once the process is initiated and this paragraph states the three permitted responses to which the Employer is limited. The consequence of selecting the response is set out in paragraph 5.	
3.	N/A	**Major change:** this is a wholly new clause for the DB 2005.	This is perhaps the most significant provision of the Contractor's Design Submission Procedure. It provides that if	

			the Employer fails to make one of the three permitted responses, then the Contractor may proceed as if the Employer has approved the Contractor's Proposal. Given a time-period of 14 days, and taking into account the likely frequency and technical complexity of the submissions, employers should realise that they must act if they wish to retain an element of control over the Works. If the Works are of a specialist nature, then the Employer may require a suitably qualified professional to be part of his team if he is to achieve the required level of scrutiny in the time available.	
4.	N/A	**Major change:** this is a wholly new clause for the DB 2005.	The paragraph provides that 'written comments' must be given when approval to proceed is not given – this is in effect a requirement for the Employer to state reasons. If the reasons were not accepted and disputed they may be the basis of an adjudication.	
5.	N/A	**Major change:** this is a wholly new clause for the DB 2005.	The action that the Contractor can adopt is stated in this paragraph. His options are limited and based entirely upon the Employer's selection (or indeed non-selection) of one of the three options.	
6.	N/A	**Major change:** this is a wholly new clause for the DB 2005.	The Contractor may not carry out any construction activity if his proposal has been rejected (marked 'C') the clause provides that the Employer will be under no obligation to pay for work carried out pursuant to a Proposal marked 'C', i.e. if the design is rejected and the Contractor does it anyway then the Employer does not have to pay.	

DB 2005 clause and reference title	WCD 1998 clause and reference title	Summary of change	Consequence of change and comments	New action required
7.	N/A	**Major change:** this is a wholly new clause for the DB 2005.	The Contractor has a short time (seven days) in which to exercise his right to object to the Employer's rejection, or comment upon his proposal. The Employer is obliged to respond in a seven-day period. Such a process may facilitate progress and design process, but is likely to be a preliminary to adjudication.	
8.	S2.2	**Major change:** the amendment changes a fundamental process or procedure of the contract compared to WCD 1998.	The paragraph supports the objection process noting the consequences of the Contractor's objection being accepted. In addition, at paragraph 8.3, a critical point is made that compliance with the Design Submission Procedure does not substitute or diminish the Contractor's obligation to design the Works in accordance with the contract – i.e. it will be no defence to an allegation that the design does not comply with the contract for the Contractor to allege that the Design Submission Procedure was complied with and that the Employer 'approved' the design under that procedure. The Contractor's Design Submission Procedure does not interfere with the Contractor's design responsibility or indeed its design liability.	
Schedule 2: Supplemental Provisions				
1. Site Manager	S3	No change: the clause repeats the text of WCD 1998.		

2. Persons named as sub-contractors in Employer's Requirements	S4.1–S4.5	Slight amendment: the text of WCD 1998 is repeated, but new defined terms are used, slight changes are made and the clause is amended to accommodate changes introduced in other clauses.	The Named Sub-Contactor regime from WCD 1998 has been all but repeated verbatim for the new form of the contract.
3. Bills of Quantities	S5, S5.1–S5. 4	Slight amendment: the clause repeats the text of WCD 1998 with very slight changes, but the effect is unchanged.	
4. Valuation of Change instructions – direct loss and/or expense – submission of estimates by the Contractor	S6.1 to S6.6	Slight amendment: the clause repeats the text of WCD 1998 but makes use of new defined terms and makes slight changes, but the effect is unchanged.	
5. Direct loss and expense – submission of estimates by Contractor	S7.1–S7.6	Slight amendment: the clause repeats the text of WCD 1998 but makes use of new defined terms and makes slight changes, but the effect is unchanged.	
Schedule 3: Insurance Options			
Insurance Option A			
New buildings – All Risks Insurance of the Works by the Contractor			
A.1 Contractor to take out and maintain a Joint Names Policy	22A.2	Slight amendment: the clause repeats the text of WCD 1998 with slight changes that improve the drafting or syntax, but the effect is unchanged.	The provisions of this Schedule have been generally updated to accommodate completion in Sections where the Parties have opted for this.
A.2 Insurance Documents – failure by Contractor to Insure	22A.2	Slight amendment: the clause repeats the text of WCD 1998 with slight changes that improve the drafting or syntax, but the effect is unchanged.	

DB 2005 clause and reference title	WCD 1998 clause and reference title	Summary of change	Consequence of change and comments	New action required
A.3 Use of Contractor's annual policy – as alternative	22A.3.1, 22A.3.2	Clause redrafted: the clause has been significantly reworded and gives additional clarity but does not give rise to new obligations.		
A.4 Loss or damage, insurance claims and Contractor's obligations	22A.4.1	Clause redrafted: the clause has been significantly reworded and gives additional clarity but does not give rise to new obligations.		
A5 Terrorism Cover – premium rate changes	22A.5.4	Slight amendment: the clause repeats the text of WCD 1998 with slight changes that improve the drafting or syntax, but the effect is unchanged.	Note that the new definition in relation to loss caused by terrorism applies, thus the loss covered by the clause is not limited to that caused by 'fire or explosion' only, but the wider 'physical loss caused by terrorism'.	
Insurance Option B				
New Buildings – All Risks Insurance of the Works by the Employer				
B.1 Employer to take out and maintain a Joint Names Policy	22B.1	Slight amendment: the clause repeats the text of WCD 1998 with slight changes that improve the drafting or syntax, but the effect is unchanged.	The provisions of this Schedule have been generally updated to accommodate completion in Sections where the Parties have opted for this.	
B.2 Evidence of Insurance	22B.2	**Significant change:** the clause repeats most of the text of the WCD 1998, but it is also amended in a way that does affect the operation of the clause.	The paragraph makes separate provision for private and local authority Employers. The procedure for private Employers follows WCD 1998. A local authority is obliged to provide a cover certificate rather than documentary evidence of the insurance. In addition, the provision permitting the Contractor then to arrange insurance cover at the expense	

184

			of the Employer will not apply if the Employer is a local authority.	
B.3 Loss or damage, insurance claims, Contractor's obligations and payment to the Employer	22B.3.1–22B.3.5	Slight amendment: the clause repeats the text of WCD 1998 with slight changes that improve the drafting or syntax, but the effect is unchanged.	Note that 22B.3.4 of WCD 1998 had a slightly wider remit. It applied to sums recoverable under the insurance policy and WCD 1998 stated that insurers were to 'pay all monies from such insurance in respect of the loss or damage' to the Employer, whereas DB 2005 provides that the Contractor authorises the insurer to pay 'all monies from such insurance'.	
Insurance Option C				
Insurance by the Employer of Existing Structures and Works in or Extensions to them				
C.1 Existing structures and contents – Joint Names Policy for Specified Perils	22C.1	Slight amendment: the clause repeats the text of WCD 1998 with very slight changes, but the effect is unchanged.		
C.2 The Works – Joint Names Policy for All Risks	22C.2	Slight amendment: the clause repeats the text of WCD 1998 with slight changes that improve the drafting or syntax, but the effect is unchanged.		
C.3 Evidence of Insurance	22C.3	**Significant change:** the clause repeats most of the text of the WCD 1998, but it is also amended in a way that does affect the operation of the clause.	The paragraph makes separate provision for private and local authority Employers. The procedure for private Employers follows WCD 1998. A local authority is obliged to provide a cover certificate rather than documentary evidence of the insurance. In addition, the provision permitting the Contractor then to arrange insurance cover at the expense of the Employer will not apply if the Employer is a local authority.	

DB 2005 clause and reference title	WCD 1998 clause and reference title	Summary of change	Consequence of change and comments	New action required
C.4 Loss or damage to Works – insurance claims and Contractor's obligations	22C.4–22C.4.4	Slight amendment: the clause repeats the text of WCD 1998 with slight changes that improve the drafting or syntax, but the effect is unchanged.		
Schedule 4: Code of Practice	Code of Practice referred to in clause 8.4.3	No change: the clause repeats the text of WCD 1998.		
Schedule 5: Third party rights				
Part 1: Third party rights for Purchasers and Tenants			The JCT's 2005 suite of standard forms is the first set of JCT contract comprehensively to address the rights that can accrue to those who ultimately occupy the buildings or finance their construction without the need to include supplements or other pro forma warranties. Users must be aware that the third party rights scheme of DB 2005 is optional and consequently the Parties must state <u>which</u> provisions they adopt in Part 2 of the Contract Particulars. A fair amount of thought is necessary to select options that are appropriate to the particular project being considered. Comparison is made to the 2001 Standard Form Agreement for Collateral Warranty for Purchasers and Tenants (Ref. MCWa/P&T). Users should note that DB 2005 creates rights for the P&T on a 'joint and several liability' basis and that the 'net contribution' clause found at 1(c) of the 2001 collateral warranty is not a term of this contract – this omission is a major change to the potential scope of the liability of the Contractor.	*Here which means as to whether the parties want to have third party involve in the contract after assigning the rights of it. II which – or Act 1999 comes into play.*

1.1	2001 Co Wa Recital C and 1(a)	Slight amendment: the clause repeats the text of the 2001 P&T collateral warranty with slight changes that improve the drafting or syntax, but the effect is unchanged.		
1.2	1(b)	Slight amendment: the clause repeats the text of the 2001 P&T collateral warranty with slight changes that improve the drafting or syntax, but the effect is unchanged.	This paragraph underpins the point that the Contractor is only liable under clause 1.1.2 if the Contract Particulars state that this is to be the case. This paragraph reinforces the point that Particulars must be completed.	
1.3	1(d)	Slight amendment: the clause repeats the text of the 2001 P&T collateral warranty with slight changes that improve the drafting or syntax, but the effect is unchanged.		
1.4	1(e)	No change: the clause repeats the 2001 P&T collateral warranty.		
2	2	Slight amendment: the clause repeats the text of the 2001 P&T collateral warranty with slight changes that improve the drafting or syntax, but the effect is unchanged.	The 'prohibited materials' themselves are not listed and instead are stated in the industry standard document produced by Ove Arup.	
3	3	No change: the clause repeats the 2001 P&T collateral warranty.	Although DB 2005 gives the P/T rights, it does not give them any formal means of directing the Contractor or of giving instructions.	
4	4	**Significant change:** the clause repeats most of the text of the 2001 P&T collateral warranty, but this is amended in a way that significantly affects the operation of the clause.	The change is that the P&T right-holder does not gain the licence (etc.) in respect of the whole of the Works – only such parts as it is the Purchaser or Tenant of. Note that, as with the 2001 P&T collateral warranty Form, P&T rights under this clause are subject to the Contractor having been fully paid (not just sums due in respect of the CDP).	

187

DB 2005 clause and reference title	WCD 1998 clause and reference title	Summary of change	Consequence of change and comments	New action required
5	5	**Significant change:** the clause automatically applies a clause that previously had to be specifically adopted by the Parties.	The clause has been redrafted – this reflects its inclusion into the main contract as opposed to publication of the clauses as a stand-alone document sufficiently flexible as to be used for an Intermediate Form or the standard JCT 1998 with Quantities as well as a WCD 1998. The reference in the Contract Particulars states that the default position in the Contract Particulars is that the professional indemnity insurance will be for the stated value in aggregate (rather than for each claim arising out of an event), unless a contrary position is stated.	State the basis on which professional indemnity insurance is required.
6	6	No change: the clause repeats the 2001 P&T collateral warranty.		
7	7	Slight amendment: the clause repeats the text of the 2001 P&T collateral warranty with slight changes that improve the drafting or syntax, but the effect is unchanged.		
8	8	Clause redrafted: the clause has been significantly reworded but does not give rise to new obligations.		
9	9	No change: the clause repeats the 2001 P&T collateral warranty.		
10	11	Clause redrafted: the clause has been significantly reworded, but does not give rise to new obligations.		

Part 2: Third party rights for a Funder			The Contractor's liability to a Funder is stated here – it is potentially far greater than that owed to the P/T. The key distinction in the scope of the rights themselves is the provision of the Funder to activate its rights to remove the Employer and take over the administration of the project (often termed activating 'step-in rights'). This is a wider set of obligations than those granted to P/T, as that clause only takes effect from practical completion. This reflects the probable involvement of the Funder before and during the construction period, in addition to the relative commercial strength of a Funder. Comparison is made to 2001 Standard Form Agreement for collateral warranty for a Funder (Ref. MCWa/F).	
1	1	**Major change:** the rights of the Parties have been changed by this amendment compared to the JCT 2001 Funder Collateral Warranty.	The change is that the net-contribution clause found in 1(a) of the 2001 collateral warranty has been deleted. This means that if the Contractor considers that others are also liable in respect of proceeding brought by a Funder, he will be obliged to consider whether to bring a separate action himself or seek to join him. Clearly this is a far more onerous process that the various, speculative and often contested calculations associated with a clause that provides that the net contributions of others are to be deducted from the Contractor's liability. Bar this deletion, the provision is identical to that in the 2001 collateral warranty.	
2	2	No change: the clause repeats the JCT 2001 Funder Collateral Warranty.		
3	3	No change: the clause repeats the JCT 2001 Funder Collateral Warranty.		

DB 2005 clause and reference title	WCD 1998 clause and reference title	Summary of change	Consequence of change and comments	New action required
4	4	No change: the clause repeats the JCT 2001 Funder Collateral Warranty.		
5	5	No change: the clause repeats the JCT 2001 Funder Collateral Warranty.		
6	6	No change: the clause repeats the JCT 2001 Funder Collateral Warranty.		
7	7	No change: the clause repeats the JCT 2001 Funder Collateral Warranty.	This clause states the consequence of the Funder stepping into the Employer's place. Simply stated, the Funder assumes all of the liability that had previously been the Employer's – specifically, all sums due or that will fall due and any obligation to pay sums due at the date of the notice. This clause provides the Contractor with a degree of protection. If the Funder activates this right, it must pay any sums outstanding to the Contractor – of course there may be some debate as to what sums are due, but the existing team of consultants are likely to be on hand to assist.	
8	8	Clause redrafted: the clause has been significantly reworded, but does not give rise to new obligations.	The Funder may make use of the CDP material in the same way and with the same limitations as the Employer. This is, however, subject to the Contractor having been fully paid (not just sums due in respect of the CDP).	
9	9	Clause redrafted: the clause has been significantly reworded, but does not give rise to new obligations.	The clause has been redrafted and makes reference to the rights stated in clause 2.38 of the main terms and conditions – this reflects its inclusion into DB 2005's main	Ensure that professional indemnity insurance is required and that the basis on which professional

			contract as opposed to being a stand-alone document capable of being used for the Intermediate Form or standard form of JCT Contract with Quantities. The reference in the Contract Particulars states that the professional indemnity insurance section must be completed to say that it will apply for the obligation to insurance to take effect, however the default position in the Contract Particulars in respect of the level of insurance is that the professional indemnity insurance will be for the stated value in aggregate (rather than for each claim arising out of an event) unless a contrary position is stated.	indemnity insurance is required is stated.
10	11	No change: the clause repeats the JCT 2001 Funder Collateral Warranty.		
11	12	No change: the clause repeats the JCT 2001 Funder Collateral Warranty.		
12	13	Slight amendment: the clause repeats the text of the JCT 2001 Funder Collateral Warranty with slight changes that improve the drafting or syntax, but the effect is unchanged.	The limitation period in respect of the P/T rights starts to count down from practical completion. Note that in respect of a Section of the Works this may be some time well before the completion of the whole of the Works. Although the note states that the limitation period should follow the method of execution, the JCT 2001 Funder Collateral Warranty simply left a blank in the text.	
13	14	No change: the clause repeats the JCT 2001 Funder Collateral Warranty.		
14		**Major change:** the amendment changes a fundamental process or procedure of the contract compared to WCD 1998.	Paragraph 14.2 had no predecessor in the 2001 Funder's Collateral Warranty, but the clause states in express terms the position that would have applied under the contract.	

191

DB 2005 clause and reference title	WCD 1998 clause and reference title	Summary of change	Consequence of change and comments	New action required
			The clause concerns the dispute resolution process following the Funder's activation of his rights. The dispute resolution process stated in the Articles will apply to the Funder once the Funder has activated its step-in rights. Thus adjudication will apply, and arbitration might apply, if the Parties have adopted it in the Articles.	
Schedule 6: Forms of Bonds				
Part 1: Advance Payment Bond	Annex 1 to Appendix 1: Terms of Bonds	Slight amendment: the clause repeats the text of WCD 1998 with slight changes that improve the drafting or syntax, but the effect is unchanged.	There is a minor deviation in the execution provisions of the Advance Payment Bond in that the Bond is now executed by a person acting as Attorney of the Surety rather than acting 'for and on behalf of the Surety'. In addition, the individual witnessing the execution must state their name and address and the Bond itself is executed as a Deed of Guarantee. Other than these formalities, the Bond and Notice of Demand are identical.	Surety's representative must now be sufficiently empowered to act as Attorney of the Surety. Ensure that witnesses confirm their identity and provide an address.
Part 2: Bond in respect of payment for off-site materials and/or goods	Bond in Respect of payment for off-site materials and/or goods	Slight amendment: the clause repeats the text of WCD 1998 with slight changes that improve the drafting or syntax, but the effect is unchanged.	Clause 3.1 has been slightly adjusted and there is also a minor deviation in the execution provisions of the Bond in that the Bond is now executed by a person acting as Attorney of the Surety rather than acting 'for and on behalf of the Surety'. The Bond itself is executed as a Deed of Guarantee. In addition, the individual witnessing the execution must state their name and address.	Surety's representative must now be sufficiently empowered to act as Attorney of the Surety. Ensure that witnesses confirm their identity and provide an address.

			Other than these formalities, the Bond and Notice of Demand are unchanged.	
Schedule 7: Fluctuations				
Fluctuations Option A Contribution, levy and fluctuations	36.1, 36.2, 36.3	Slight amendment: the clause repeats the text of WCD 1998 but makes use of new defined terms and makes slight changes, but the effect is unchanged.	The text now refers directly to the Contract Sum rather than 'the prices contained in the Contract Sum and the Contract Sum Analysis' as it did in WCD 1998, however this change is not thought to have consequence. The term 'Base Date' is defined by reference to the Contract Particulars. The only other change is the updating of references to legislation. The provisions of the Fluctuations Option A are the same as clauses 36.1–36.2 of WCD 1998. The relevant change is that A1.2 now refers to the *Industrial Training Act* 1982 rather than the *Industrial Training Act* 1964.	
Fluctuations Option B Labour and materials cost and tax fluctuations	37.1–37.8	Slight amendment: the clause repeats the text of WCD 1998 but makes use of new defined terms and makes slight changes, but the effect is unchanged.	The text now refers directly to the Contract Sum rather than 'the prices contained in the Contract Sum and the Contract Sum Analysis' as it did in WCD 1998, however this change is not thought to have consequence. The term 'Base Date' is defined by reference to the Contract Particulars. The only other change is the updating of references to legislation. The provisions of the Fluctuations Option B are the same as clauses 37.1–37.8 of WCD 1998. The relevant change is that B2.2 now refers to the *Industrial Training Act* 1982 rather than the *Industrial Training Act* 1964.	

DB 2005 clause and reference title	WCD 1998 clause and reference title	Summary of change	Consequence of change and comments	New action required
Fluctuations Option C Formula adjustment	38.1–38.6	Slight amendment: the clause repeats the text of WCD 1998 but makes use of new defined terms and makes slight changes, but the effect is unchanged.		

Contract clause headings and numbering structure from the *Design and Build Contract*, Joint Contracts Tribunal Limited, 2005, Sweet and Maxwell, © The Joint Contracts Tribunal Limited 2005, are reproduced here with permission.

Table of destinations: CD 1998 and DB 2005

This table of destinations follows the structure of the old JCT 1998 and states the new clause reference in the 2005 form.

CD 1998	DB 2005	CD 1998	DB 2005
Parties to Agreement	Parties to Agreement	1.4	3.19
First recital	First recital	1.5	1.7
Second recital	Second recital	1.6	1.5
Third recital	Third recital	1.7	1.5 and 1.11
Fourth recital	Fourth recital	1.8	1.8
Fifth recital		1.9	1.6
Sixth recital		2.1	2.1
Article 1	Article 1	2.2	1.3
Article 2	Article 2	2.3	2.10
Article 3	Article 3	2.4.1	2.12. 2.13 and 2.14.2
Article 4	Article 4	2.4.2	2.14.1
Article 5	Article 7	2.4.3	2.13
Article 6A	Article 8	2.5.1	2.17
Article 6B	Article 9	2.5.2	2.17
Article 7.1	Article 5	2.5.3	2.17
Article 7.2	Article 6	3	4.3
Article 8	Fifth recital	4.1.1	3.5
Attestation	Attestation	4.1.2	3.6
1.1	1.2 and 1.3	4.2	3.8
1.2	1.3	4.3	3.7
1.3	1.1	5.1	2.7

CD 1998	DB 2005
5.2	2.7
5.3	2.8 and Schedule 1, para 1
5.4	2.7
5.5	2.37
5.6	2.7 and 2.38
6.1.1	2.1
6.1.2	2.15
6.1.3	2.16
6.2	2.18
6.3	2.15
6A.1	3.18
6A.2	3.18
6A.3	3.18
6A.4	3.19
6A.5	3.18
7	2.9
8.1.1	2.2
8.1.2	2.2
8.1.3	2.1
8.2	2.2
8.3	3.12
8.4	3.13
8.5	3.14
8.6	2.2
9.1	2.19
9.2	2.20

CD 1998	DB 2005
10	3.2
11	3.1 and 3.4
12.1	5.1
12.2	3.9
12.3	3.11
12.4	5.2
12.5.1	5.4
12.5.2	5.4
12.5.3	5.4
12.5.4	5.5
12.5.5	5.6
12.5.6	5.7
12.6	5.3
13	4.1
14.1	
14.2	4.4
14.3	4.4
15.1	2.21 and 2.23
15.2	4.15
15.3	2.22
16.1	2.27
16.2	2.35
16.3	2.35
16.4	2.36
17.1	2.30
17.1.1	2.31

CD 1998	DB 2005
17.1.2	2.32
17.1.3	2.33
17.1.4	2.34
18.1	7.1 and 7.2
18.2	7.2
18.2.1	3.3 and 3.4
18.3	3.4
19	
20.1	6.1
20.2	6.2
20.3	6.3
21.1.1	6.4
21.1.2	6.5
21.2	6.5
21.3	6.6
22.1	6.7
22.2	6.8
22.3	6.9
22A.1	
22A.2	Schedule 3, paras A1 and A2
22A.3.1	Schedule 3, para A3
22A.4	Schedule 3, para A4
22A.5	6.10
22A.5.4	Schedule 3, para A5
22B.1	Schedule 3, para B1
22B.2	Schedule 3, para B2

CD 1998	DB 2005
22B.3	Schedule 3, para B3
22B.4.1	6.10
22B.4.2	6.10
22B.4.3	6.10
22C.1	Schedule 3, para C1
22C.1A	
22C.2	Schedule 3, para C2
22C.3	Schedule 3, para C3
22C.4	Schedule 3, para C4
22C.5	6.10
22D	
22FC.1	6.13
22FC.2	6.14
22FC.3	6.15
22FC.4	
22FC.5	6.16
23.1	2.3
23.1.2	2.4
23.2	3.10
23.3.1	2.3
23.3.2	2.5
23.3.3	2.5
24.1	2.28
24.2	2.29
25.1	
25.2.2	2.24
25.3.1	2.25.1 and 2.25.2

JCT 2005 Design and Build Contract (DB 2005)

CD 1998	DB 2005
25.3.1.3	2.25.3
25.3.1.4	2.25.3
25.3.2	2.25.1 and 2.25.4
25.3.3	2.25.5
25.3.4	2.25.6
25.3.5	2.25.6
25.4	2.26
26.1	4.19
26.2	4.20.1
26.3	4.21
26.4	4.22
27.1	8.2
27.2.1	8.4
27.2.2	8.4
27.2.3	8.2 and 8.4
27.2.4	8.2
27.3.1	8.1
27.3.2	8.5
27.4	8.6
27.5.1	8.5
27.5.4	8.5
27.6	8.7
27.7.1	8.8
27.8	8.3
28.1	8.9
28.2.2	8.9.2

CD 1998	DB 2005
28.3	8.10
28.4	
28.5	8.3
28A.1	8.11
28A.2	8.12
29.1	2.6
29.2	2.6
29.3	
30A	4.5
30.1.1	4.6 and 4.7
30.1.2	4.6 and 4.8
30.2A.1	4.13
30.2A.4	4.13
30.2B	4.14
30.3.1	4.9
30.3.2	4.9
30.3.3	4.10.3
30.3.4	4.10.4
30.3.5	4.10.5
30.3.6	4.10.1
30.3.7	4.10.6
30.3.8	4.11
30.4.1	4.17
30.4.2	4.16
30.4.3	4.10.2
30.5	4.12

CD 1998	DB 2005	CD 1998	DB 2005
30.5.3	4.2	36.2	Schedule 7, Part 1
30.6	4.12	36.3	Schedule 7, Part 1
30.7		36.4	
30.8	1.9	36.5	
30.9	1.10	36.6	
31.1		36.7	
31.2	4.5	37	Schedule 7, Part 2
31.3		38	Schedule 7, Part 3
31.4		39A	9.2
31.5		39B.1.1	9.4
31.6		39B.2	9.5
31.7		39B.3	9.6
31.8		39B.4	9.7
31.9		39B.5	9.8
31.10		39B.6	9.3
31.11		39C	
31.12		Supplementary Provisions	
31.13		S1	
31.14		S2.1	Schedule 1, para 1
32		S2.2	Schedule 1, para 2
33		S2.3	
34.1	3.15	S3	Schedule 2, para 1
34.2	3.16	S4	Schedule 2, para 2
34.3.1	3.17	S5	Schedule 2, para 3
35	4.18 and Schedule 7	S6	Schedule 2, para 4
36.1	Schedule 7, Part 1	S7	Schedule 2, para 5

4   JCT Intermediate Building Contract 2005

Introduction

The Intermediate Building Contract of 2005 (IC 2005) replaces the JCT Intermediate Form of Building Contract of 1998 (IFC 1998). The intermediate form has been widely used in practice and is seen as one of the most significant of the JCT's standard forms of contract due to its massive uptake.

Background to the intermediate forms

As a procurement tool, the 'intermediate' forms of contract have sought to bridge the gap between the full JCT Contract (with or without quantities) and the more simple 'minor works' forms. Users of the IFC 1998 forms evidently thought that it achieved a suitable balance of the detail, formality and completeness that is required for larger projects and the accessibility, comprehensibility and brevity required for more modest projects. That the balance had been achieved in that form was indicated by the substantial numbers of the contract that were purchased.

When one considers the scale of the projects that are likely to be procured under IC 2005, users are presented with a limited number of alternative forms to choose from. The Third Edition of the New Engineering Contract (NEC) includes a 'Short Form'. The Short Form, like all NEC contracts, takes a fundamentally different approach to the procurement model of the JCT contracts. Procurement under NEC can be highly effective if applied correctly. However, success is often directly proportional to attention given to the project management process of the contract in terms of the time and resource devoted to

applying the NEC mechanisms for advance warnings and compensation events at the time they appear. This administrative overhead is an essential element of the NEC method of working and its proponents insist that better project management, in terms of cost certainty and quality, are the result. Irrespective of this, such an investment may be considered disproportionate to the value of the Works being undertaken – particularly if the value of the project is 'intermediate'.

The JCT Drafting Committee appears to have been mindful of the fact that such a widely used and popular form of contract ought not be changed wantonly. The guidance notes no longer state an upper value limit on the projects that the contract is considered to be appropriate for. This change is, reportedly, to take account of the trend for the IFC to be used towards the top of the value range stated, so as not unduly to discourage users from taking up the form if the value is moderately in excess of the stated range but the form is otherwise appropriate. In addition, IC 2005 is buttressed by a newly drafted sub-contract for use by Contractors. This back-to-back Intermediate Sub-Contract Agreement is to be published in a number of formats to support a variety of sub-contractors, including design responsibilities.

Purpose and structure of this chapter

This chapter is divided into two parts. The first part briefly introduces the principal changes and notes concepts that have remained unchanged. The second part forms the main body of the chapter, providing a clause by clause analysis of the new IC 2005 with the aim of identifying and explaining the differences between IFC 1998 and IC 2005.

The second part of this chapter has been set out in a tabular format to aid use as a point of reference when Contract Documents are being

JCT 2005 Intermediate Building
Contract (IC 2005)

prepared and when the contract is used in practice on site. Reading from left to right, the columns are:

- Column One: states the number and name of the IC 2005 clause;
- Column Two: states the number and name (where there is one) of the IFC 1998 clause;
- Column Three: a brief statement summarising any change and its nature. Standard descriptions are used to indicate the degree of change at a glance;
- Column Four: examines the specific significance of any changes, including a consideration of the consequences of the differences for the operation of the contract. As such, this column is the heart of the book;
- Column Five: states new actions that the Parties must perform and, where relevant, provides guidance on what this means in practice.

The third and final part is a table of destinations plotting that follows the IFC 98 format and states the location of the clause in IC 2005.

Base documents

In this chapter, the IC 2005 has been compared to IFC 1998 incorporating Amendments 1: 1999, 2: 2000, 3: 2001, 4: 2002, 5: 2003.

Summary of the changes in IC 2005

Users familiar with the IFC 1998 will find that there are a limited number of substantive changes that are specific to the intermediate form of contract. Although the contract has been revised (and, when taken cumulatively, the changes have significance), the changes are quite generic and reflect the general changes to the 2005 suite of contracts rather than a fundamental recasting of the structure or obligations. Those generic 'JCT 2005' changes are generally aimed at reducing uncertainty and providing a clear course of action to Parties

as they carry out the Works and overcome the practical difficulties associated with building, and they do mean that IC 2005 is a moderate improvement on IFC 1998.

Significant changes

Restructuring and redrafting

The layout of the contract has been improved. Although the new contract adopts the same format as the other new JCT contracts, the change is especially noticeable in IC 2005, as the layout no longer follows IFC 1998's 'two-column' format where each of the A4 pages of the contract had text arranged in two vertical columns.

The clauses themselves are grouped into sections and the clauses are given clear titles and the layout is generally less cluttered. This is more than just a housekeeping exercise: it should result in the document being better understood and applied by its users in practice. Positive changes to the drafting and greater economy of expression are perhaps slightly obscured by some changes in the names of various familiar terms. For example, the 'Defects Liability Period' has been renamed the 'Rectification Period' and 'extensions of time' are now dealt with under the general heading of 'Adjustments to the Completion Date' – such changes in terminology are pointed out in the second part of this chapter. Certain ambiguous clauses have also been redrafted in the interests of clarity.

Changes of substance

Contract Particulars
All standard form contracts require the individual details that tailor the standard form to the particular project to be stated in or appended to the document that is ultimately to be signed. IC 2005 follows the format of the other contracts in the JCT's 2005 release by seeking to place most of the contract-specific information at the start of the

contract in what are termed the 'Contract Particulars'. The IFC 1998 required much the same information to be stated in the Appendix towards the end of the contract.

The contract has also moved towards stating a default position on many of the options that Parties may select, or that reflect regulations to be accommodated by the Parties. The effect is that if the Parties want to apply a different period for the rectification of defects or an alternative rate of interest to that provided, they will need to amend the form to deviate from the default.

Dispute resolution

The dispute resolution processes have been comprehensively revised in line with the changes introduced in all of the other JCT standard forms. Most significantly, the IFC 1998 stated that arbitration would be the default dispute resolution method, for all disputes – the IC 2005 reverses this and, unless the Parties specifically select arbitration in the Contract Particulars, then it will not be the default mechanism for disputes under the contract. This means that the Parties may use the courts to assert their rights in addition to seeking the temporarily binding decisions of an Adjudicator.

The adjudication process itself has also been changed substantially: the new standard form provides that the Parties may identify an Adjudicator Nominating Body to propose an Adjudicator or may select an Adjudicator for the contract in the Contract Particulars. The new form also makes better provision for the situation where the Parties have failed to complete the section or left it blank – in that circumstance a Party requiring a matter to be adjudicated may select any of the listed nominating bodies to make a nomination. The JCT Adjudication Rules have been dropped in favour of the rules of the statutory *Scheme for Construction Contracts (England and Wales) Regulations* 1998 ('the Scheme'); it is thought that this reflects experience with, and the authority that now supports, the Scheme.

The contract also has a new clause that suggests that the Parties consider the mediation of disputes. The mediation process has found much favour in the courts, due to its ability to resolve disputes at an early stage with a minimum level of cost to the Parties and a limited amount of court time. The mediation process can be of particular benefit on projects where the value of the project, and therefore possibly the value of the dispute, are moderate in proportion to the potential costs. Thus mediation may be especially well suited to resolving disputes under IC 2005.

Termination

The termination clauses have been redrafted and the change runs deeper than the change in terminology from the phrase 'determination'. The definitions of the various acts of 'insolvency' have been updated, but more significantly a notice of termination must be given to an Insolvent Party to terminate the contract on that ground. IFC 1998 did not require such a notice and automatically deemed the termination to be effective upon the instance of the insolvency.

A new right for either Party to terminate in the event of a two-month suspension is stated. If the contract is terminated by either Party under this clause, or if the contract has been terminated as a result of the Employer's default, then there is a new obligation on the Contractor to prepare an account to determine the sums due rather than wait for the Contractor to take action.

Sectional completion

The new form incorporates provisions that will enable the Parties to have the work carried out and completed in Sections. Parties wishing to make use of IFC 1998 and have the work carried out in Sections would have needed to make the additional purchase of the JCT Sectional Completion Supplement 1998.

Changes to assist in execution

The requirement to state the company number of the contracting Party should assist in providing an absolute way of identifying the contracting entity – which can be especially difficult when contracting with a Party with a number of operating divisions or subsidiaries at various locations.

Electronic communications

IFC 1998 made provision for Electronic Data Interchange – this option has been deleted from the standard form of IC 2005. The replacement is a facility for the Parties to list which notices and correspondence required by the contract might legitimately be given by electronic means. If this is required, they must complete the section in the Contract Particulars. The Parties might conceivably provide that all communication and notices required under the contract could be sent electronically.

The risk of such change is that it downgrades the apparent significance of documents that have very serious consequences if they are not dealt with in an appropriate manner. The other risk is that the correspondence may be vague or an important notice might be appended to correspondence on unrelated matters.

It would seem that professionalism would safeguard against such consequences; provided that the staff drafting the electronic correspondence do it correctly and state its purpose clearly, and that the staff receiving the correspondence are alive to the potential that each email might have, then quick and clear communication can ensure that the contract is operated effectively.

Many projects are to all intents and purposes managed in this way, with instructions and payment certificates being first issued by email often in PDF format on the same day that the hard copies are sent in the post.

Daywork

Clause 5.4 of IC 2005 has clarified the limited number of instances when Daywork is the appropriate basis for charging and provides the requirement that vouchers be provided by way of evidence of the time spent.

Advance payment bond

The new form provides that an advance payment bond shall be provided by the Contractor, if the Employer provides an advance payment.

Significant items that have not changed

Certain features of the IFC 1998 have been retained in IC 2005 despite the fact that similar features have been deleted or amended in other JCT standard forms being released in 2005.

Named Sub-Contractors

The contract retains a method of appointing a Named Sub-Contractor this is found in Schedule 2. The processes set out in the contract are substantially unchanged in comparison to IFC 2005. Other JCT 2005 issue contracts have dispensed with naming provisions.

Retention

Retention under IC 2005 is set at 5%, which is the same level as it was set under IFC 1998, despite the fact that it has been stated at reduced levels in SBC 2005 and DB 2005.

JCT 2005 Intermediate Building Contract (IC 2005)

IC 2005 clause and reference title	IFC 1998 clause and reference title	Summary of change	Consequence of change and comments	New actions
Articles of Agreement	'This Agreement…'	**Significant amendment:** the clause repeats most of the text of the IFC 1998, but this is amended in a way that does affect the operation of the clause.	The Parties are asked to give a company number, where there is one. The company number provides an additional reference point that can easily be checked on the Companies House website (www.companieshouse.com). Company numbers are unique to the particular company and do not change, unlike the name or registered address of the company which may change. Stating the company number gives additional clarity as to the identity of the Parties.	Limited companies should state their company number.
Recitals				
First Recital *(the Works)*	First Recital	**Significant change:** the clause has been redrafted and this affects its operation.	Unlike IFC 1998, IC 2005 does not state that the Works are carried out 'under the direction of the Architect / the Contract Administrator'.	
Second Recital *(the Contract Drawings)*	First Recital	**Significant amendment:** the clause repeats much of the text of the JCT 1998, but also provides for new process or action from the Parties.	The drawings are to be listed and/or numbered and identified here (either by writing in the numbers or referring to a list of the numbered drawings). The requirement for both Parties to mark the drawings with a signature or initials is now stated here.	The list of the drawings is to be annexed to the contract. The parties are to sign the drawings.

			If carried out correctly, these exercises should benefit the Parties in terms of providing greater certainty as to the required scope of work.	
Third Recital *(documents provided by the Employer)*	First Recital	Significant change: the clause repeats most of the text of the IFC 1998 but it is also amended in a way that does affect the operation of the clause.	This Recital states the documents that may, depending on the project, be provided by the Employer such as Bills of Quantities, Specification and Work Schedules. The change is that IC 2005 improves the provisions in respect of the documents to be given by the Employer when a Named Sub-Contractor is to be used. It does this by specifying the terms and conditions of sub-contract. This is to be the Intermediate Named Sub-Contract Tender and Agreement, a new form published by JCT for 2005. In addition, the Recital requires the transfer of significant documents such as Bills of Quantities, Specification and Work Schedules but also documents such as the Sub-Contract Invitation to Tender and the Tender itself.	
Fourth Recital *(the Priced Document)*	Second Recital	Clause redrafted: the clause has been significantly reworded, but does not give rise to new obligations.		
Fifth Recital *(Construction Industry Scheme)*	5.6 and Appendix Ref to 5.6	Clause redrafted: the clause has been significantly reworded but does not give rise to new obligations.	The status of the Employer for the purposes of the Construction Industry Scheme is to be stated in the Contract Particulars .	
Sixth Recital *(Information Release Schedule)*	Fourth Recital	Slight amendment: the clause repeats the text of the IFC 1998 with slight changes that improve the drafting or syntax, but the effect is unchanged.		

JCT 2005 Intermediate Building Contract (IC 2005)

IC 2005 clause and reference title	IFC 1998 clause and reference title	Summary of change	Consequence of change and comments	New actions
Seventh Recital (*CDM Regulations*)	Third Recital	Slight amendment: the clause repeats the text of the IFC 1998 with slight changes that improve the drafting or syntax, but the effect is unchanged.		
Eighth Recital (*Sections*)	Article 7	Clause redrafted: the clause has been significantly reworded, but does not give rise to new obligations.	If the project is to be carried out in Sections, then this should be noted in detail in the Contract Particulars.	
The Articles			*Generally:* The Articles for IC 2005 have been edited and it appears that contemporary words have been favoured where possible.	
Article 1: Contractor's Obligations	Article 1	Clause redrafted: the clause has been significantly reworded, but does not give rise to new obligations.	Statement that 'The Contractor shall carry out and complete the Works in accordance with the Contract Documents'. The explicit statement that this obligation is undertaken for a consideration has been removed, but a valid contract would nevertheless be created as the exchange of obligations is apparent and adequately dealt with by subsequent Articles.	
Article 2: Contract Sum	Article 2	Slight amendment: the clause repeats the text of IFC 1998 with slight changes that improve the drafting or syntax, but the effect is unchanged.		
Article 3: Architect/ Contract Administrator	Article 3	Slight amendment: the clause repeats the text of IFC 1998 but makes use of new defined terms and makes slight changes, but the effect is unchanged.	The Article identifies the Architect/ Contract Administrator (A/CA), but the Article itself no longer provides processes for the replacement of the consultants. These have been relocated to clause 3.4 of the main conditions.	

Article 4: Quantity Surveyor	Article 4	Slight amendment: the clause repeats the text of IFC 1998 but makes use of new defined terms and makes slight changes, but the effect is unchanged.	Identifies the Quantity Surveyor, but the Article itself no longer provides processes for the replacement of the consultants. These have been relocated to the main body of the contract at clause 3.4.	
Article 5: Planning Supervisor	Article 5	**Major change:** the amendment changes a fundamental process or procedure of the contract compared to IFC 1998.	Whilst the process is not changed, the amended clause establishes the default position more effectively and avoids the possibility of there not being a Planning Supervisor. Thus the Architect/Contract Administrator is to be the Planning Supervisor unless there is a specific appointment to the contrary. The Employer should therefore ensure that the provisions of his agreement with the Architect/Contract Administrator include planning supervision services.	State the identity of the Planning Supervisor unless the Parties wish this to be the A/CA. The Employer should agree with the A/CA that he is providing planning supervision services.
Article 6: Principal Contractor	Article 6	Clause redrafted: the clause has been significantly reworded and gives additional clarity, but does not give rise to new obligations.	As with IFC 1998, the default position is that the Contractor will be the Principal Contractor for the purposes of CDM. There is a slight difference in that provision is made to state the identity of someone else if that other person is to be the Principal Contractor.	
Article 7: Adjudication	Article 8	No change: the clause repeats the text of IFC 1998.		
Article 8: Arbitration	Article 9A	**Major change:** the amendment changes a fundamental process or procedure of the contract compared to IFC 1998.	The most significant change found in the Articles. IC 2005 does not provide for arbitration unless the Parties specifically select it. Arbitration can still be adopted as the default dispute resolution process amending the Contract Particulars to reflect this preference.	The merits of arbitration should be considered and if it is decided that it is required, the Contract Particulars must be amended.

209

IC 2005 clause and reference title	IFC 1998 clause and reference title	Summary of change	Consequence of change and comments	New actions
Article 9: Legal proceedings	Article 9B	**Significant amendment:** the clause has been significantly reworded, but does not give rise to new obligations.		
Contract Particulars	Appendix		*General:* The Contract Particulars are significantly expanded in comparison to the 1998 form and they now provide the project-specific information that was previously found in the Appendix at the end of the IFC 1998.	
Part 1 General				
Fifth Recital and clause 4.4	N/A	**New definition:** this clause provides a new definition of a term or of a process.	This is an explicit reference to the Construction Industry Scheme (CIS) which deals with the basis on which a Contractor is to be taxed.	The Employer and/or his advisors should be aware of the requirements of the CIS to avoid falling foul of them. Advice from the Employer's accountants or finance department may be required to ascertain whether the Employer is deemed to be a Contractor for these purposes.
Seventh Recital	Appendix reference to Seventh Recital	No change: the clause repeats the text of IFC 1998.		
Eighth Recital	Sectional Completion Supplement 1998	**Significant change:** the clause has been significantly reworded and accommodates the incorporation of the Sectional Completion provisions, but does not give rise to new obligations.	The IC 2005 provides for the option of the Works being divided into, and provided, in Sections. The necessary changes are found throughout the document. If sectional completion were to be a requirement of an IFC 1998 contract, then this would necessitate the purchase and incorporation of the separate JCT Sectional Completion Supplement.	

Article 8	Appendix reference to Articles 9A and 9B	**Major change:** the amendment changes a fundamental process or procedure of the contract compared to IFC 1998.	IFC 1998 contained a clause that required that Parties in dispute would take that dispute to an Arbitrator rather than to the courts; the courts would stay any action in the courts in deference to the Party's agreement to adopt arbitration in accordance with the *Arbitration Act* 1996. The new default position for IC 2005 is diametrically opposite to that stated in IFC 1998 and is that arbitration will not apply to disputes arising under the contract and disputes will be resolved through the courts. If the Parties want arbitration to be the dispute resolution method for the contract instead of litigation, then this must be stated in the Contract Particulars and the default position changed. If the Parties decide that arbitration would be appropriate for a specific dispute even though they have not selected arbitration as the default mechanism, then they are free to agree to this on a case-by-case basis. Note also that adjudication will apply to construction contracts irrespective of the selection made.	The merits of arbitration should be considered, and if it is required it should be stated in this part of the Contract Particulars.
1.1	Appendix reference to 8.3	No change: the clause repeats the text of IFC 1998.		
1.1	Appendix reference to date for completion and Sectional Completion Supplement 1998	Clause redrafted: the clause has been significantly reworded and accommodates the incorporation of the Sectional Completion provisions, but does not give rise to new obligations.		
1.7	1.13	**New definition:** this clause provides a new definition of a term or of a process.	The clause provides for the Parties to state an address and fax number for receipt of notices; it also provides for a default in the event that the Parties fail to complete this section. In such a case, the address stated in the commencement section on IC page 1 will be the address for service of notices.	Both the Employer and the Contractor should include an address and a contact name for the service of contractual or other notices. It is in the interests of both Parties for the

211

JCT 2005 Intermediate Building Contract (IC 2005)

IC 2005 clause and reference title	IFC 1998 clause and reference title	Summary of change	Consequence of change and comments	New actions
			Although this is a pragmatic clause, there is the risk that Parties might send the notice to the wrong place – particularly if the address identified is not the registered office or 'principal place of business' of the company.	address to be one where if a notice were received, it would be acted upon immediately.
1.8	1.16	**Major change:** the amendment changes a fundamental process or procedure of the contract compared to IFC 1998.	The IFC 1998 provision for an optional Electronic Data Interchange has been removed. The Parties are now free to agree that the listed communications may be validly issued by electronic means. The Contract Particulars provide a space for the Parties to make a list of the communication, documents and notices etc., that can be validly sent electronically. If the Parties do not provide such a list, then the contract default position is that communications must be in writing and the consequence would be that electronic documents would have no formal status under the contract. The JCT has not suggested which of the communications required by the contract might sensibly be transferred electronically. At one end of the sliding scale would be a notice of termination, which it is suggested would not be appropriate for electronic notice; and at the other end the lower-level enquiries and requests for information where it would be practical. Difficulty would arise with the matters in the middle – for example, perhaps it is appropriate for withholding notices to be sent by email, but inappropriate for instructions to be sent by email.	Decide whether or not to consent to valid notice or communication by email, etc. If this is selected then the specific items of correspondence and the form of communication must be set down in this section of the Contract Particulars for this preference to take effect.

2.4	Appendix reference to 2.1	Clause redrafted: the clause has been significantly reworded and accommodates the incorporation of the Sectional Completion provisions, but does not give rise to new obligations.	The Contract Particulars require the statement as to the Date of Possession. The redraft accommodates possession of Sections being given at different points in times.	
2.5 and 2.20.3	Appendix reference to deferment	Clause redrafted: the clause has been significantly reworded and accommodates the incorporation of the Sectional Completion provisions but does not give rise to new obligations.	The change to the drafting of the deferment provision is the inclusion of a provision to permit deferment of possession of sections. Note that the Parties must state at this point in the Contract Particulars whether or not there is to be any right of deferment.	
2.23.2	Appendix reference to 2.23.2	Clause redrafted: the clause has been significantly reworded and accommodates the incorporation of the Sectional Completion provisions but does not give rise to new obligations.	The provision for liquidated damages is unchanged save that it is extended to accommodate liquidated damages being paid on late sectional completion.	
2.29	Sectional Completion Supplement Clause	Clause redrafted: the clause has been significantly reworded and accommodates the incorporation of the Sectional Completion provisions but does not give rise to new obligations.		
2.30	Appendix reference to Defects Liability Period	Clause redrafted: the clause has been significantly reworded and accommodates the incorporation of the Sectional Completion provisions but does not give rise to new obligations.	The name of the defects liability period has been changed to 'Rectification Period'. In addition, provision has been made for the statement of separate Rectification Periods (which may be of differing duration) associated with a project that will have Sectional Completion. The default Rectification Period for either a Section or the whole of the Works is six months.	

IC 2005 clause and reference title	IFC 1998 clause and reference title	Summary of change	Consequence of change and comments	New actions
4.5	Appendix reference to 4.2(b)	No change: the clause repeats the text of IFC 1998.	The default position is that there is no advance payment and therefore if needed the clause requires specific activation in the Contract Particulars. Note that this provision does not apply if the Contractor is a local authority.	It is important that the Employer and the Contractor complete the section governing reimbursement of the advance payment. If the payment is reimbursed by the Employer too quickly, it could have an adverse effect on the Contractor's cashflow. If reimbursed too slowly, the Employer's cashflow could be affected.
4.5	Appendix reference to 4.2(b)	**Major change:** the amendment changes a fundamental process or procedure of the contract compared to IFC 1998.	If the Parties agree to an advance payment then an advance payment bond must be provided by the Contractor as a default unless the Contract Particulars state here that the Bond is not required. IFC 1998 made the provision of an Advance Payment Bond optional rather than mandatory. Advance Payments do not apply if the Employer is a local authority, thus this provision does not apply nor is it necessary if the Employer is a local authority.	
4.6.1	Appendix reference to 4.2(a)	Slight amendment: the clause repeats the text of IFC 1998 with slight changes that improve the drafting or syntax, but the effect is unchanged.		
4.7.1	Appendix reference to 4.2.1	**New definition:** this clause provides a new definition of a term or of a process.	IFC 1998 placed the percentage to be paid prior to the start of the Rectification Period in the main body of the contract, whereas IC 2005 states this sum in the Contract Particulars.	

4.9		**New definition:** this clause provides a new definition of a term or of a process.		
4.12.4	Appendix reference to 4.2.1(c)1	Slight amendment: the clause repeats the text of IFC 1998 with slight changes that improve the drafting or syntax, but the effect is unchanged.		
4.12.5	Appendix reference to 4.2.1(c)2	Slight amendment: the clause repeats the text of IFC 1998 with slight changes that improve the drafting or syntax, but the effect is unchanged.		
4.15 and Schedule 4	Appendix reference to 4.9(a) and 4.9(b)	Slight amendment: the clause repeats the text of IFC 1998 with slight changes that improve the drafting or syntax, but the effect is unchanged.	The default position is that the Fluctuations Option under Schedule 4 applies. IC 2005 includes a footnote that points out that the Fluctuations Options are not appropriate to all circumstances, particularly if there is likely to be a short construction period.	If the Parties require a more absolute lump sum contract, then consideration should be given to the deletion of all of the fluctuation clauses.
6.4.1.2	Appendix reference to 6.2.1	Slight amendment: the clause repeats the text of IFC 1998 with slight changes that improve the drafting or syntax, but the effect is unchanged.	Although the reference is virtually identical to IFC 1998, the purpose of the clause is now stated in terms of being in respect of injury to persons or property.	
6.5.1	Appendix reference to 6.2.4	Slight amendment: the clause repeats the text of IFC 1998 with slight changes that improve the drafting or syntax, but the effect is unchanged.	In addition to the familiar text of IFC 1998 there is also a note making the express statement that insurance of this kind is only necessary if the Parties specifically state they may require it and state a minimum value at the time of contract.	
6.7 and Schedule 1	Appendix reference to 6.3.1	Slight amendment: the clause repeats the text of IFC 1998 with slight changes that improve the drafting or syntax, but the effect is unchanged.		
6.7 and Schedule 1 Insurance Option	Appendix reference to 6.3A.1, 6.3B.1, 6.3C.2	**Major change:** the amendment changes a fundamental process or procedure of the contract compared to IFC 1998.	IC 2005 provides that coverage of 15% is the default sum required in respect of professional fees connected with insurance events.	

215

JCT 2005 Intermediate Building Contract (IC 2005)

IC 2005 clause and reference title	IFC 1998 clause and reference title	Summary of change	Consequence of change and comments	New actions
			The provision of a default position in the absence of any change to the Contract Particulars is a major change. This was not provided in IFC 1998.	Users who require cover for professional fees should check that the default percentage is correct for their particular project.
6.7 and Schedule 1 Insurance Option	Appendix reference to 6.3A.3.1	Slight amendment: the clause repeats the text of IFC 1998 with slight changes that improve the drafting or syntax, but the effect is unchanged.		
6.11	6.3FC.1	No change: the clause repeats the text of IFC 1998.		
6.14	6.3FC.5	**Significant change:** the clause repeats most of the text of IFC 1998 but this is amended in a way that does effect the operation of the clause.	If the Contract Particulars do not state a preference, then the default position will apply and the Contractor will be obliged to bear the cost of compliance with amendments or revision to the Joint Fire Code.	
8.9.2	7.9.2	**Major change:** the amendment changes a fundamental process or procedure of the contract compared to IFC 1998.	The Parties may now state how long must pass before the contract can be terminated on the grounds of suspension. The default time period for suspension has been increased from one month to two months for IC 2005.	Both Parties should consider whether the two-month default period is acceptable. Employers should consider whether or not they can permit suspension for such a period; and Contractors whether they can maintain an appropriate level of mobilisation without work and without the payment that would be associated with it.

8.11.1.1–8.11.1.5	7.13.1	**Major change:** the amendment changes a fundamental process or procedure of the contract compared to IFC 1998.	The Parties may now state how long must pass before the contract can be terminated on the grounds of suspension due to force majeure or government action. The default time period for suspension has been decreased from three months to two months for IC 2005.	
9.2.1	Appendix reference to 9A.2	**Major change:** the amendment changes a fundamental process or procedure of the contract compared to IFC 1998.	Adjudicator appointment under IC 2005 is different from IFC 1998 in a number of respects. IC 2005 provides that the Parties may nominate a specific individual in the Contract Particulars to act as Adjudicator – this was not provided for in IFC 1998. If the Contract Particulars have not been completed, then a referring Party may select any of the listed Adjudicator Nominating Bodies (ANB). Under IFC 1998 the position was different and the contract defaulted to the RIBA as ANB.	Some thought should be given to the pros and cons of naming an Adjudicator. If the Parties name an experienced and impartial Adjudicator, they may be more confident of an acceptable decision should the need arise. Conversely, where an Adjudicator Nomination Body is chosen, the likelihood of obtaining an Adjudicator with the relevant skills and experience increases.
9.4.1	9B.1	Slight amendment: the clause repeats the text of IFC 1998 with slight changes that improve the drafting or syntax, but the effect is unchanged.	The remit of the appointer under IC 2005 is slightly wider (or at least the role is more explicit) than that under IFC 1998 in that the appointment of a replacement Arbitrator is also covered. As such, this is a more complete draft that provides a resolution when a specific difficulty is encountered.	
Part 2 Collateral Warranties	None	**Major change:** the rights of the Parties have been changed by this amendment compared to IFC 1998.	The third party rights and collateral warranty regime for IC 2005 is new and has no precedent in IFC 1998, but it is also quite different to the mechanisms in other JCT 2005 editions such as SBC/Q 2005 and DB 2005. Clause 7 of IC 2005 presents a range of collateral warranties that may be required of the Contractor and from his sub-contractors. The collateral warranties may	Employers must decide if they require collateral warranties and whether or not they need to secure collateral warranties for Funders and the ultimate occupiers.

IC 2005 clause and reference title	IFC 1998 clause and reference title	Summary of change	Consequence of change and comments	New actions
			be in favour of the Employer, a Funder or Purchasers and Tenants. It is crucial that if these rights are required, this is indicated at this point in the Contract Particulars. The default position is that there will not be an obligation to provide any collateral warranties. Unlike the larger 2005 JCT contracts, IC 2005 does not present an option for the Parties to make use of the *Contracts (Rights of Third Parties) Act* 1999. Similarly, IC 2005 does not include the text of the warranties in its Schedules, whereas these documents are found in the larger new forms. The warranties themselves will have to be purchased. Although this is not a new action, it is a departure from other 2005 forms of JCT contract. Some Employees may wish to continue to rely upon their own standard warranties.	
Attestation	Attestation page headed 'as witness the hands of the parties hereto'	Slight amendment: The clause repeats the text of the IFC 1998 with slight changes that improve the drafting or syntax, but the effect is unchanged.	The attestation process is unchanged but the words used are more modern and the notes in the IC 2005 itself are clearer than the IFC 1998. Like the IFC 1998, the IC 2005 provides attestation forms for limited companies and individuals. The guidance note points out that the attestation may not be correct for housing associations, partnerships and foreign companies but this is not stated in the terms of the contract itself.	

CONDITIONS

Section 1: Definitions and Interpretation			Definitions that are not amended or where the change does not have material effect are not noted. Several definitions that had been found in the text of IFC 1998 are now located, or at least referenced, in this part of IC 2005. The definitions also serve as something approaching an index, as if the term is defined in the text of the main contract this is noted and the clause reference is given.	
Definitions				
Activity Schedule	8.3 Activity Schedule	Clause redrafted: the clause has been significantly reworded but does not give rise to new obligations.		
Agreement	Articles or Articles of Agreement	Slight amendment: the clause repeats the text of IFC 1998 but makes use of new defined terms and makes slight changes, but the effect is unchanged.	The new defined term incorporates a reference to the new Contract Particulars section of the contract.	
Business Day	N/A	**Major change:** the amendment changes a fundamental process or procedure of the contract compared to IFC 1998.	The definition 'Business Day' is new, however the reckoning of periods of days (clause 1.5 of IC 2005 and clause 1.14 of IFC 1998) and definition of 'Public Holiday' are unchanged. This is relevant because many of the time-periods stated in the contract make reference to 'days', not 'Business Days'. For example, clause 2.15.2, concerning divergence from Statutory Requirements, makes no reference to 'Business Days'. The point is that users should apply the appropriate period of time in order to ensure that activities are carried out or replies to notices are issued in the correct time-period.	

IC 2005 clause and reference title	IFC 1998 clause and reference title	Summary of change	Consequence of change and comments	New actions
Contract Particulars	Appendix	**Significant amendment:** this is a new clause that accommodates changes introduced in other clauses.	This is a new definition describing the Contract Particulars and also noting that they are completed by the Parties.	
Contractor's Persons	N/A	**Significant amendment:** this is a new clause that accommodates changes introduced in other clauses.	A useful catch-all definition that describes who the Contractor is responsible for.	
Employer's Persons	N/A	**Significant amendment:** this is a new clause that accommodates changes introduced in other clauses.	A catch-all definition that describes who the Employer is responsible for. Note that the Architect/Contract Administrator and the Quantity Surveyor are expressly excluded from the scope of this definition.	
Finance Agreement Funder	N/A	**Significant amendment:** this is a new clause that provides a new definition of a term or of a process, but does not cause a change to the rights or obligations of the Parties.	New definitions to accommodate the provision of collateral warranties to third parties.	
Health and Safety Plan	8.3 Health and Safety Plan	**Significant amendment:** the clause repeats most of the text of the IFC 1998, but this is amended in a way that does affect the operation of the clause.	The second half of the IFC 1998 dealt with the Contractor's obligation to continue to provide the Health and Safety Plan before construction began and reflect changes during the Works. This obligation is now effectively subsumed within clause 3.18.2 of IC 2005.	
Information Release Schedule	8.3 Information Release Schedule	Clause redrafted: the clause has been significantly reworded, but does not give rise to new obligations.	The clause now refers to the provisions of the Sixth Recital. The previous version of the contract made reference to the potential to change the Information Release Schedule – this is now found in clause 2.10 of IC 2005 instead.	

Listed Items	4.2.1(c)	**Significant change:** the clause repeats most of the text of the IFC 1998, but it is also amended in a way that does affect the operation of the clause.	IFC 1998 stated that for an item to be a 'Listed Item' it had to be identified as such and listed before delivery, but this is no longer stated as being necessary. The IFC 1998 version also required full compliance with a number of other clauses for the item to fall within the definition 'Listed Item'. The present version of the contract states no such requirement, but deals with the issue at clause 4.12 by providing the Contractor with no right to payment unless the Conditions are met.
Provisional Sum	8.3 Provisional sum	**Significant change:** the clause repeats most of the text of the IFC 1998, but it is also amended in a way that does affect the operation of the clause.	Although the definition of Provisional Sum in relation to Contract Documents with Contract Bills is unchanged the definition has been significantly expanded and now covers projects where the Contract Documents do not include Contract Bills. As such, this is a more thorough statement of what is generally meant by the term 'provisional sum'. This should reduce the prospect of difficulties of interpretation and application of the clause.
Purchaser Tenant	N/A	**New definition:** this clause provides a new definition of a term or of a process.	These new definitions support the new regime for providing rights for Purchasers and Tenants.
Rectification Period	Defects Liability Period	**Significant amendment:** the clause has been significantly reworded and gives additional clarity, but does not give rise to new obligations.	A change to the name of the period for correcting defects.

IC 2005 clause and reference title	IFC 1998 clause and reference title	Summary of change	Consequence of change and comments	New actions
Interpretation				
1.2 Reference to clauses etc.	8.1	Slight amendment: the clause repeats the text of IFC 1998 but makes use of new defined terms and makes slight changes, but the effect is unchanged.		
1.3 Articles etc. to be read as a whole	1.3, 8.2	**Significant change:** this is a new clause that provides a new definition of a term or of a process, but does not cause a change to the rights or obligations of the Parties.	Clause 1.3 provides a hierarchy for the elements of the contract. This new hierarchy places the Agreement and the Conditions at the top of the list and the provides that the Contract Bills, Specification, Work Schedules or their contents will not override them.	Users should take time to ensure that the Contract Particulars, especially, are correct, in view of the hierarchy of the obligations.
1.4 Headings, reference to persons, legislation etc.	N/A	**Significant change:** this is a new clause that provides a new definition of a term or of a process, but does not cause a change to the rights or obligations of the Parties.	This is a new clause that provides the basis on which the terms of the contract are to be interpreted. The clause does not actually create new obligations but adds value by clarifying the interpretation of the contract.	
1.5 Reckoning periods of days	1.14	No change: the clause repeats the text of IFC 1998.		
1.6 Contracts (Rights of Third Parties) Act 1999	1.17	No change: the clause repeats the text of IFC 1998.		
1.7 Giving or service of notices or other documents	1.13	Clause redrafted: the clause has been significantly reworded and gives additional clarity, but does not give rise to new obligations.		
1.8 Electronic communications	1.16	**Major change:** the amendment changes a fundamental process or procedure of the contract compared to IFC 1998.	The provision for an Electronic Data Interchange in IFC 1998 has been deleted.	

			The Contract Particulars provide a space for the Parties to make a list of the documents that can be validly served electronically. The effect of the clause is that the Parties may state that email and other forms of electronic communication will be a valid form of communication and may proscribe its validity for certain kinds of communication in accordance with their preference. If the Contract Particulars contain no such list, then electronic communication will not be a valid method of serving a notice that the contract states must be served by first class post or by actual delivery.	Complete the Contract Particulars if electronic communication is considered satisfactory or appropriate for key contract processes.
1.9 Issue of Architect/ Contract Administrator's certificates	1.9	Slight amendment: the clause repeats the text of IFC 1998 with slight changes that improve the drafting or syntax, but the effect is unchanged.		
1.10 Effect of Final Certificate	4.7.1	**Significant change:** the clause repeats most of the text of the IFC 1998 but this is amended in a way that does affect the operation of the clause.	Much of the original text of IFC 1998 is repeated verbatim. There are a number of respects in which the Final Certificate is not conclusive and IC 2005 now provides that fraud will remove the conclusivity of the certificate. This provision makes express a term that was hitherto implied. Note that the fraud alluded to may be that of the Architect/Contract Administrator in assembling the certificate, but could equally be the fraud of a Contractor pertaining to the granting of the Final Certificate.	
1.11 Effect of certificates other than Final Certificate	4.8	No change: the clause repeats the text of IFC 1998.		
1.12 Applicable Law	1.15	Clause redrafted: the clause has been significantly reworded, but does not give rise to new obligations.	Fewer words are used to ensure the same provision.	

JCT 2005 Intermediate Building Contract (IC 2005)

IC 2005 clause and reference title	IFC 1998 clause and reference title	Summary of change	Consequence of change and comments	New actions
Section 2: Carrying out the Works				
Contractor's Obligations				
2.1 General obligations	1.1, 5.1	**Significant change:** the clause repeats most of the text of the IFC 1998, but this is amended in a way that does affect the operation of the clause.	The clause states the Contractor's core obligations. There is a slight deviation from the text of the IFC 1998 in that it is open to interpretation whether the phrase 'in compliance with the Contract Documents' found in IC 2005 is wider than 'in accordance with the Contract Documents'. In addition, the express obligation to carry out the Works in compliance with the Statutory Requirements has been relocated to this clause from clause 5.1 of IFC 1998.	
2.2 Materials, goods and workmanship	1.1, 1.2	**Significant change:** the clause repeats most of the text of the IFC 1998, but this is amended in a way that does affect the operation of the clause.	As with IFC 1998 the standards to be applied are those stated in the Contract Documents. The standard of the reasonable satisfaction of the Architect/ Contract Administrator has been retained. The clause now provides for the situation where no standard has been stated in the Contract Documents. In such cases, the standards of goods and workmanship are to be 'of a standard appropriate to the Works' – this is admittedly a standard that will be difficult to judge objectively. Clause 2.2.2 also introduces a well-intentioned principle that will be difficult to apply. The requirement is in respect of the training of staff by the Contractor with the obligation stated that the Contractor take 'all reasonable steps to encourage'	Contractors are to encourage staff to train under the Construction Skills Certification Scheme.

			such training. The idea is entirely laudable and restates the amendment introduced by 1.1 of Amendment 5 dated 1 July 2003; however it may well prove difficult to enforce.	
2.3 Fees and charges	5.1	Clause redrafted: the clause has been significantly reworded, but does not give rise to new obligations.	The clause requires the Contractor to pay out in the event of fees or taxes being levied upon the Works, but grants the Contractor the right to recover the outlay from the Employer and for the outlay to be added to the Contract Sum. The Contractor is not permitted to either recover the sums nor add them to the Contract Sum if the Priced Documents or Specification provided that the Contract Sum was to have included them. The clause is an improved version of that found in the second paragraph of clause 5.1 of IFC 1998.	Fees and taxes that are likely to be levied on the Works should be considered and allocated. If they are not allocated to the Contractor in the Specification or Priced Document, then the Employer will ultimately be responsible for reimbursing the Contractor.
Possession				
2.4 Date of Possession – progress	2.1	Clause redrafted: the clause has been significantly reworded and accommodates the incorporation of the Sectional Completion Supplement, but does not give rise to new obligations.	Although reworded, the clause retains the same obligations as its predecessor in IFC 1998. Note that the express reference to the extension of time has been removed, perhaps because it is otiose. The date of possession is now expressed as the point at which the Contractor is obliged to 'begin the construction of the Works' – this obligation is not effective until possession has been granted; other obligations concerning non-construction activity or prefabrication may become effective before this point.	
2.5 Deferment of possession	2.2	Slight amendment: the clause repeats the text of IFC 1998 with slight changes that improve the drafting or syntax, but the effect is unchanged.		

225

IC 2005 clause and reference title	IFC 1998 clause and reference title	Summary of change	Consequence of change and comments	New actions
2.6 Early use by Employer	2.1	Slight amendment: the clause repeats the text of IFC 1998 with slight changes that improve the drafting or syntax, but the effect is unchanged.		
2.7 Work not forming part of the Contract	3.11	Clause redrafted: the clause has been significantly reworded and gives additional clarity but does not give rise to new obligations.		
Supply of Documents, Setting Out etc.				
2.8 Contract Documents	1.6, 1.8	**Significant change**: the clause repeats most of the text of the IFC 1998 but it is also amended in a way that does affect the operation of the clause.	The clause is an amalgamation of two familiar IFC 1998 provisions and is in substantially the same form. Thus it covers both the provision of documents to the Contractor and what the Contractor can do with them and the limitations on his use of them. The first distinction is that it is now an express requirement that the documents to be provided to the Contractor by the Architect/Contract Administrator are to be provided *immediately*. IFC 1998 did not state a time-period and a term would have to be implied to the effect that the documents would be provided within a reasonable time. Secondly, it applies if Pricing Option B is being used, in which case the requirement on the Employer to retain the Contract Documents found in IFC 1998 now applies to the Priced Documents too.	

2.9 Levels and setting out of the Works	3.9	Slight amendment: the clause repeats the text of IFC 1998 with slight changes that improve the drafting or syntax, but the effect is unchanged.		
2.10 Construction information	1.7.1	Slight amendment: the clause repeats the text of IFC 1998 with slight changes that improve the drafting or syntax, but the effect is unchanged.	The only change of any interest is that the clause is now headed 'Construction Information' in IC 2005, whereas it used to be headed 'Information Release Schedule' in clause 1.7.1 of IFC 1998.	
2.11 Further drawings, details and instructions	1.7.2	Slight amendment: the clause repeats the text of IFC 1998 with slight changes that improve the drafting or syntax, but the effect is unchanged.	The clause has been set out in three separate subclauses rather than a single paragraph and the result is greater clarity.	
Errors, Inconsistencies and Divergences				
2.12 Bills of Quantities	1.5, 1.4	Slight amendment: the clause repeats the text of IFC 1998 with slight changes that improve the drafting or syntax, but the effect is unchanged.	Although there is little here that is new, the clause is an amalgam of several IFC 1998 provisions. Clause 2.12.1 of IC 2005 restates the provisions of clause 1.5 of IFC 1998. Clause 2.12.2 of IC 2005 repeats the provisions of clause 1.4 of IFC 1998 in respect of the correction of such deficient information pertaining to a Provisional Sum for defined work as may be required by the Standard Method of Measurement.	
2.13 Instructions on errors, omissions and inconsistencies	1.4	**Significant change:** the clause repeats most of the text of the IFC 1998 but this is amended in a way that does affect the operation of the clause.	The obligation upon the Architect/Contract Administrator to issue instructions to correct errors etc., is stated here. An express obligation has been added that makes it incumbent upon the Contractor to give immediate written notice of errors (etc.) to the Architect/Contract Administrator.	Contractor to advise Architect/Contract Administrator of errors immediately.

JCT 2005 Intermediate Building Contract (IC 2005)

IC 2005 clause and reference title	IFC 1998 clause and reference title	Summary of change	Consequence of change and comments	New actions
2.14 Instructions – additions to Contract Sum, exceptions	3.6.1	Clause redrafted: the clause has been significantly reworded, but does not give rise to new obligations.	The clause ensures that instructions resolving errors that constitute a Variation are priced in accordance with the Variations provisions.	
2.15 Divergences from Statutory Requirements	5.2, 5.3	**Significant change:** the clause repeats most of the text of the IFC 1998 but it is also amended in a way that does affect the operation of the clause.	The clause has been redrafted and expanded in comparison to clauses 5.2 and 5.3 of IFC 1998. Under IC 2005, as with IFC 1998, the Contractor will generally not be liable to the Employer if he does what he is required to, either by the Contract Documents or by an instruction, even if this is not in accordance with Statutory Requirements. However, IC 2005 also adds a reference to divergence from Statutory Requirements in the drawings that the Architect/Contract Administrator provides either when giving levels or when providing accurately dimensioned drawings to enable the Contractor to set out the Works. Whilst the changes to the clause are significant, the Contractor must still give written notice to the Architect/Contract Administrator of any divergence immediately and in writing and he will not gain the benefit of the clause if he does not comply with this obligation to give notice. The duty of the Architect/Contract Administrator in the event of the discovery of divergence has also been changed. Clause 2.15.2 of IC 2005 provides that once the divergence is known to the Architect/Contract Administrator (either via notice from the Contractor or from its own efforts or those of the Employer), it	The Architect/Contract Administrator is to respond to discovery of divergence with an instruction within seven days of being made or becoming aware of it.

			has seven days in which to issue instructions ' in that regard' whereas IFC 1998 did not state a period of time for a response.	
2.16 Emergency compliance with Statutory Requirements	5.4		Clause redrafted: the clause has been significantly reworded and gives additional clarity, but does not give rise to new obligations.	
Unfixed Materials and Goods – property, risk etc.				
2.17 Materials and goods on site	1.10		Slight amendment: the clause repeats the text of IFC 1998 with slight changes that improve the drafting or syntax, but the effect is unchanged.	IFC 1998 referred to the Employer making payment for the unfixed materials under a 'payment certificate', whereas IC 2005 refers to the Employer having paid for the unfixed materials under 'any Interim Certificate'.	Employers and Architect/Contract Administrators should ensure that all payments are made in accordance with the contract terms under an Interim Certificate – this is to avoid confusion and the risk of paying twice for the same item.
2.18 Materials and goods off site	1.11		Slight amendment: the clause repeats the text of IFC 1998 with slight changes that improve the drafting or syntax, but the effect is unchanged.	IFC 1998 referred to the Employer making payment for the unfixed materials under a 'payment certificate', whereas IC 2005 refers to the Employer having paid for the unfixed materials under 'any Interim Certificate'.	Employers and Architect/Contract Administrators should ensure that all payments are made in accordance with the contract terms under an Interim Certificate – this is to avoid confusion and the risk of paying twice for the same item.

IC 2005 clause and reference title	IFC 1998 clause and reference title	Summary of change	Consequence of change and comments	New actions
Adjustment of Completion Date				
2.19 Notice of delay – extensions	2.3, 2.5	Slight amendment: the text of IFC 1998 is repeated, new defined terms are used and the clause now accommodates the incorporation of the Sectional Completion provisions; notwithstanding this, the effect of the clause is unchanged.	Such small changes as have been made reflect the creation of new defined terms such as 'Relevant Events' and the incorporation of the regime for completion in Sections. In addition, the new clause 2.19.5 ensures that arguments concerning the applicability of the clause to further extensions of time are reduced.	Users should consider updating references in relevant documentation, e.g. notices, certificates, awards etc. from stating an 'extension of time' to stating an 'adjustment to Completion Date' so as to align with the new terminology.
2.20 Relevant Events	2.4	**Major change:** the amendment changes a fundamental process or procedure of the contract compared to IFC 1998.	The clause has been redrafted, and to a certain extend codified. It gains from the use of the defined term 'Relevant Event' but relies upon a catch-all provision in the form of clause 2.20.6, which accounts for the failures of the Employer and his team (it is in similar form to the catch-all clause added to IFC 1998 by Amendment 4 of 2002). This wider clause has meant that a number of the Relevant Events found in IFC 1998 no longer need to be stated separately in IC 2005. The following Relevant Events from IFC 1998 have been subsumed within the scope of IC 2005's clause 2.20.6: ● 2.4.7.1(Information Release Schedule); ● 2.4.7.2 (lack of architect's drawings); ● 2.4.8 (Employer or others working on site); ● 2.4.9 (Employer's default in supply of equipment); ● 2.4.17 (compliance/non-compliance with CDM).	

				A number of events that would have resulted in an extension of time have been deleted; these are: • 2.4.10 (Contractor's failure to secure labour); • 2.4.11 (Contractor's failure to secure materials).	
Practical Completion, Lateness and Liquidated Damages					
2.21 Practical completion and certificates	2.9, Sectional Completion Supplement	Slight amendment: the text of IFC 1998 is repeated, new defined terms are used and the clause now accommodates the incorporation of the Sectional Completion provisions; notwithstanding this, the effect of the clause is unchanged.	The new defined terms should be noted – these are 'Practical Completion Certificate' and 'Section Completion Certificate'.		
2.22 Certificate of non-completion	2.6	Slight amendment: the clause repeats the text of IFC 1998 with slight changes that improve the drafting or syntax, but the effect is unchanged.			
2.23 Liquidated damages for non-completion	2.7	**Significant change:** the clause repeats most of the text of the IFC 1998, but this is amended in a way that does affect the operation of the clause.	The conditions for the retention of liquidated damages remain the same as in IFC 1998. The Employer is still entitled to either recover the sums from the Contractor as a debt, or alternatively withhold it from sums that are to become due. There is a change in that whichever of the options the Employer selects, it must inform the Contractor of this by notice (clause 2.7.1 of IFC 1998 provided that communication advising the Contractor that liquidated damages would be recovered as a debt to be merely in writing).		

IC 2005 clause and reference title	IFC 1998 clause and reference title	Summary of change	Consequence of change and comments	New actions
			In addition, the clause now expressly states that the Employer may retain or withhold a sum less than the amount that has been pre-agreed. Note also that the term used in IC 2005 is 'liquidated damages' rather than 'liquidated and ascertained damages'. It does not appear that this change is intended to have material consequence, however the concept of damages that are not 'ascertained' would appear to suggest a lower degree of association with the loss that the sums are intended to represent a genuine pre-estimate of. This might invite an attempt to prove a lack of *any* degree of pre-estimation in court in order to demonstrate that the sums are in fact penalties and as such should not be enforced by a court.	Employers and their advisors should ensure that any sum stated as liquidated damages is sufficiently grounded on the facts of the likely losses.
2.24 Repayment of liquidated damages	2.8	Slight amendment: the clause repeats the text of IFC 1998 with slight changes that improve the drafting or syntax, but the effect is unchanged.	Clause 2.8 of IFC 1998 provided that subsequent non-completion certificates be taken into account when considering whether sums were to be repaid if a non-completion certificate were to be revoked. The reference to clause 2.22 of IC 2005 achieves a similar function. The clause is all but the same.	
Partial Possession by Employer				
2.25 Contractor's consent	2.11	Slight amendment: the clause repeats the text of IFC 1998 with slight changes that improve the drafting or syntax, but the effect is unchanged.		

2.26 Practical completion date	2.11	Slight amendment: the clause repeats the text of IFC 1998 with slight changes that improve the drafting or syntax, but the effect is unchanged.	IC 2005 gives this matter its own clause – it was previously a sub-clause of IFC 1998 as clause 2.11.	
2.27 Defects etc. – Relevant Part	2.11	Slight amendment: the clause repeats the text of IFC 1998 with slight changes that improve the drafting or syntax, but the effect is unchanged.	IC 2005 gives this matter its own clause – it was previously a sub-clause of IFC 1998 clause 2.11.	
2.28 Insurance – Relevant Part	2.11	Slight amendment: the clause repeats the text of IFC 1998 with slight changes that improve the drafting or syntax, but the effect is unchanged.	IC 2005 gives this matter its own clause – it was previously a sub-clause of IFC 1998 clause 2.11.	
2.29 Liquidated damages – Relevant Part	2.11	Slight amendment: the clause repeats the text of IFC 1998 with slight changes that improve the drafting or syntax, but the effect is unchanged.	IC 2005 gives this matter its own clause – it was previously a sub-clause of IFC 1998 clause 2.11.	
Defects				
2.30 Rectification	2.10	**Significant change:** the clause repeats most of the text of the IFC 1998, but this is amended in a way that does affect the operation of the clause.	The phrase 'defects liability period' is replaced by the new defined term Rectification Period'. In addition, the old standard provision found in several JCT contracts including IFC 1998 that pre-Practical Completion frost is at the risk of the Contractor is no longer explicitly stated in this clause as a defect.	
2.31 Certificate of making good	2.10	Slight amendment: the clause repeats the text of IFC 1998 with slight changes that improve the drafting or syntax, but the effect is unchanged.	IC 2005 gives this matter its own clause – it was previously a sub-clause of IFC 1998 clause 2.10.	

233

JCT 2005 Intermediate Building Contract (IC 2005)

IC 2005 clause and reference title	IFC 1998 clause and reference title	Summary of change	Consequence of change and comments	New actions
Section 3: Control of the Works				
Access and Representatives				
3.1 Access for Architect/ Contract Administrator	N/A	**Major change:** the rights of the Parties have been changed by this amendment compared to IFC 1998.	Notwithstanding the Contractor's possession of the site, the Architect/ Contract Administrator has access to the Works – this is an express statement of a right that was formerly implied. It is notable that the provision providing access extends to 'the Works and elsewhere to any work which is being prepared for or utilised in the Works' – this would clearly cover an off-site manufacturing facility where specific elements are being made or prefabricated.	
3.2 Person-in-charge	3.4	Slight amendment: the clause repeats the text of IFC 1998 with slight changes that improve the drafting or syntax, but the effect is unchanged.		
3.3 Clerk of Works	3.10	No change: the clause essentially repeats the text of IFC 1998.	There is a very minor difference in this clause as the Clerk of Works acts 'as inspector' in IC 2005, whereas he acts 'as an inspector' in IFC 1998.	
3.4 Replacement of Architect/Contract Administrator or Quantity Surveyor	Article 3, Article 4	**Major change:** the amendment changes a fundamental process or procedure of the contract compared to IFC 1998.	The Contractor has a limited right to object to the individual that the Employer proposes as the replacement Architect/ Contract Administrator or as replacement Quantity Surveyor. There are a number of differences in the new form compared to IFC 1998.	

			The rights do not apply if the Employer is a local authority and it is seeking to appoint one of its officials. The IC 2005 refers differences on the suitability of the proposed replacement to the dispute resolution process. The process does not result in the professional being selected by the tribunal – instead, if the proposed replacement is proved to be sufficiently objectionable it requires the Employer to nominate an acceptable replacement. If the replacement is not acceptable to the Contractor, then the solution is go thorough the dispute resolution process once again. Clause 3.4.2 provides a useful clarification to the effect that subsequent appointees are permitted to review decisions (or change their minds) to the same extent and with the same consequences as their predecessors would have been able to.	
Sub-Letting				
3.5 Consent to sub-letting	3.2	Slight amendment: the clause repeats the text of IFC 1998 with slight changes that improve the drafting or syntax, but the effect is unchanged.	Note that clause 3.2 of IFC 1998 and its several sub clauses has been broken into separate clauses.	
3.6 Conditions of sub-letting	3.2, 3.2.1, 3.2.3	**Significant change:** the clause repeats most of the text of the IFC 1998, but this is amended in a way that does affect the operation of the clause.	Other than the following points the clause makes the same provision as IFC 1998. There is a new sub-clause at 3.6.4 that requires the incorporation of a term into a sub-contract to require a collateral warranty from the sub-contractor. IC 2005 provides that the sub-contractor may not remove unfixed materials without the consent of the Contractor. It then goes on to say that the Contractor's consent cannot be unreasonably delayed or withheld – in both these respects clause 3.6.2.1 is the same as IFC 1998. The	Contractors should insert into the relevant sub-contracts the obligations that require the provision of warranties. Employers and/or their advisors should, where possible, inspect sub-contracts in advance of their execution to ensure that the required term has been included.

IC 2005 clause and reference title	IFC 1998 clause and reference title	Summary of change	Consequence of change and comments	New actions
			distinction is that clause 3.6.3 of IC 2005 makes the express provision that the consent of the Architect/Contract Administrator is required before the Contractor can give this consent to a sub-contractor.	
3.7 Named Sub-Contractors	3.3.1	Slight amendment: the clause repeats the text of IFC 1998 with slight changes that improve the drafting or syntax, but the effect is unchanged.	The provisions relating to Named Sub-Contractors have been relocated to Schedule 2 of IC 2005. IC 2005 also requires that the Contractor engage the Named Sub-Contractor on the 'Intermediate Named Sub-Contract Agreement IC Sub/NAM/A'.	
Architect/Contract Administrator's Instructions				
3.8 Compliance with instructions	3.5.1	No change: the clause essentially repeats the text of IFC 1998.		
3.9 Non-compliance with instructions	3.5.1	**Significant change:** the clause repeats most of the text of the IFC 1998 but this is amended in a way that does effect the operation of the clause.	Compared to IFC 1998, IC 2005 states a more restrictive basis for the recovery of the Employer's costs if the Contractor fails to comply with an instruction. Under IC 2005 the Employer may only recover 'additional costs incurred by the Employer in connection with such employment', whereas IFC 1998 enabled the Employer to deduct 'all costs incurred' to employ and pay for the instruction to be executed by others. Unlike IFC 1998, IC 2005 does not provide that the sum may be recovered as a debt. IC 2005 provides that there should be an appropriate deduction to the Contract	

			Sum for work carried out by others. This too is new.	
3.10 Provisions empowering instructions	3.5.2	Slight amendment: the clause repeats the text of IFC 1998 with slight changes that improve the drafting or syntax, but the effect is unchanged.		
3.11 Instructions requiring Variations	3.6	Clause redrafted: the clause has been significantly reworded but does not give rise to new obligations.		
3.12 Postponement of work	3.15	Slight amendment: the clause repeats the text of IFC 1998 with slight changes that improve the drafting or syntax, but the effect is unchanged.		
3.13 Instructions on Provisional Sums	3.8	Clause redrafted: the clause has been significantly reworded, but does not give rise to new obligations.		
3.14 Inspections – tests	3.12	Slight amendment: the clause repeats the text of IFC 1998 but makes use of new defined terms and makes slight changes, but the effect is unchanged.		
3.15 Work not in accordance with the Contract	3.13.1	Slight amendment: the clause repeats the text of IFC 1998 but makes use of new defined terms and makes slight changes, but the effect is unchanged.		
3.16 Instructions as to removal of work etc.	3.14.1, 3.14.2	Slight amendment: the clause repeats the text of IFC 1998 but makes use of new defined terms and makes slight changes but the effect is unchanged.		
3.17 Exclusion of persons from the Works	N/A	**Major change:** the rights of the Parties have been changed by this amendment compared to IFC 1998.	This is a new clause that provides the express power to the Architect/Contract Administrator to exclude persons from the site and as such reflects a power stated in many other JCT contracts.	

IC 2005 clause and reference title	IFC 1998 clause and reference title	Summary of change	Consequence of change and comments	New actions
			Whilst the power is limited by the requirement that it is not exercised unreasonably or vexatiously, it should not be thought that it is purely a 'health and safety' provision – even if this may be its most obvious application.	
CDM Regulations				
3.18 Undertakings to comply	5.7.1, 5.7.2, 5.7.4	Slight amendment: the clause repeats the text of IFC 1998 with slight changes that improve the drafting or syntax, but the effect is unchanged.		
3.19 Appointment of successors	1.12, 5.3	Clause redrafted: the clause has been significantly reworded and gives additional clarity, but does not give rise to new obligations.	The redrafted clause requires the Employer to give the Contractor written notice if it replaces the Planning Supervisor or Principal Contractor.	If the Employer replaces the Planning Supervisor or Principal Contractor he is to give the Contractor written notice of this.
Section 4: Payment				
Contract Sum and Adjustment				
4.1 Work included in Contract Sum	1.2	Slight amendment: the clause repeats the text of IFC 1998 with slight changes that improve the drafting or syntax, but the effect is unchanged.		
4.2 Adjustment only under the Conditions	4.1	Slight amendment: the clause repeats the text of IFC 1998 with slight changes that improve the drafting or syntax, but the effect is unchanged.		

Certificates and Payments				
4.3 VAT	5.5	**Significant change:** the clause repeats most of the text of the IFC 1998, but it is also amended in a way that does affect the operation of the clause.	The only change of significance is the reference in the last line of clause 4.3.2 to the sums that the Contactor may claim from the Employer in the event that the Employer is no longer eligible to pay VAT. Under IFC 1998 the sums had to relate to the tax paid on "goods and services which contribute exclusively to the Works'. Under IC 2005 the redrafted clause has a less restrictive scope in that it need no longer be exclusive to the specific Works, but is more exacting in that the sums will only be reimbursed if the sums paid cannot be recovered as a consequence of the exemption. The consequence is that if a Contractor has procured otherwise taxable goods for a variety of projects, this will not prevent him recovering sums equivalent to the tax already paid in respect of goods supplied to a tax-exempt Employer. The Contractor will make this recovery from the tax-exempt Employer.	
4.4 Construction Industry Scheme (CIS)	4A, 5.6	Slight amendment: the clause repeats the text of IFC 1998 with slight changes that improve the drafting or syntax, but the effect is unchanged.		
4.5 Advance payment	4.2(b)	**Major change:** the amendment changes a fundamental process or procedure of the contract compared to IFC 1998.	IC 2005 requires that an advance payment bond is required in each instance that an advance payment is to be made. The Contractor's provision of the advance payment bond is stated to be a precondition to the Employer's obligation to make the advance payment.	An advance payment bond is now mandatory in the circumstances where an advance payment is to be made to the Contractor If the Parties opt to provide an advance

239

IC 2005 clause and reference title	IFC 1998 clause and reference title	Summary of change	Consequence of change and comments	New actions
			IFC 1998 required the Parties to state the requirement for an advance payment bond on a case-by-case basis. Note that provision for an advance payment is itelf optional.	payment, then the Employer is not obliged to make that advance payment unless and until the Contractor has provided the bond.
4.6 Interim Certificates and Valuations	4.2(a), 4.2(c)	Slight amendment: the clause repeats the text of IFC 1998 with slight changes that improve the drafting or syntax, but the effect is unchanged.	The text is substantially unchanged, however the old clause 4.2 has been rearranged: the IC 2005 version divides up that clause stating the certificate process, amounts due and payment separately. This clause 4.6 concerns the timing of Interim Certificates, Interim Valuations and Contractor's applications for payment. The clause updates the language favouring a more contemporary word selection by replacing the IFC 1998 phrase 'without prejudice to the obligation' with 'without affecting ... the obligation'.	
4.7 Amounts due in Interim Certificates	4.2(d)	Slight amendment: the clause repeats the text of IFC 1998 with slight changes that improve the drafting or syntax, but the effect is unchanged.		
4.7.1	4.2.1, 4.2.1(a), 4.2.1(c)	Clause redrafted: the clause has been significantly reworded (and gives additional clarity), but does not give rise to new obligations.	Note that the amount to be paid (and by inference the amount to be retained) is stated in the Contract Particulars – the unamended contract provides for a 95% payment and a retention at 5%.	
4.7.2	4.2.2	Clause redrafted: the clause has been significantly reworded (and gives additional clarity), but does not give rise to new obligations.	The clause is redrafted in that a number of the defined terms and names of the items referred to in IFC 1998 have been changed in IC 2005.	

4.7.3	4.2.2	**Major change:** the amendment changes a fundamental process or procedure of the contract compared to IFC 1998.	In addition, the clause now makes explicit provision for the deduction of sums that have been incurred as a result of the Contractor's non-compliance with instructions.	
4.8 Interim Certificates – payment				
4.8.1	4.2(a)	**Significant change:** the clause repeats most of the text of the IFC 1998, but it is also amended in a way that does affect the operation of the clause.	The significant change relates to the deletion of a number of provisions in IFC 1998 to accommodate specific circumstances such as stage payment – the removal of this material clarifies the text. Clause 4.8.1 improves on IFC 1998 as it refers directly to payment pursuant to an Interim Certificate (clause 4.2(a) of IFC 1998 referred to payment against certificates).	
4.8.2	4.2.3(a)	No change: the clause repeats the text of IFC 1998.		
4.8.3	4.2.3(b)	Slight amendment: the clause repeats the text of IFC 1998 with slight changes that improve the drafting or syntax, but the effect is unchanged.		
4.8.4	4.2.3(c)	Clause redrafted: the clause has been significantly reworded and gives additional clarity, but does not give rise to new obligations.	The redraft does give a marginal increase in clarity of the clause. The clause provides that the Employer may only pay a sum different to the sum stated in the Interim Certificate if he issues a certificate under clause 4.8.2, in which case he must pay that sum. The Employer may then only pay a sum different to that in his statement under 4.8.2 if he submits a withholding notice under 4.8.3 in which case he may validly pay the sum stated in the certificate under 4.8.3.	

241

IC 2005 clause and reference title	IFC 1998 clause and reference title	Summary of change	Consequence of change and comments	New actions
4.8.5	4.2(a)	Clause redrafted: the clause has been significantly reworded and gives additional clarity but does not give rise to new obligations.		
4.9 Interim payment on practical completion	4.3	**Significant change:** the clause repeats most of the text of the IFC 1998, but it is also amended in a way that does affect the operation of the clause.	IC 2005 uses the same process for the interim payment due on practical completion as it does for the payment of a standard Interim Payment Certificate. The processes for payment statements and withholding statements for the Interim Certificate on practical completion are incorporated by a cross-reference to those clauses. The change is that IFC 1998 repeated the text on the interim payment mechanism in clause 4.3(a)–(c). The comments in respect of the payment / withholding process at clause 4.8 (above) also apply to this clause.	Reference should be made to the Contract Particulars to check whether or not the standard practical completion payment percentage has been changed from the default 97.5%.
4.10 Interest in percentage withheld	4.4	No change: the clause repeats the text of IFC 1998.		
4.11 Contractor's right of suspension	4.4A	**Significant change:** the clause repeats most of the text of the IFC 1998, but it is also amended in a way that does affect the operation of the clause.	The last sentence of the clause 4.4A of IFC 1998 that ensured that the Contractor exercising its right to suspend did not amount to a suspension that might lead to termination or alternatively a failure to proceed diligently with the Works. It is suggested that the consequence of this clause is to reinforce the existing position that if the Contractor's reasons for the suspension were not valid, then it might well be in breach of its other duties and this might itself give rise to a right of termination.	Contractors should act with even greater caution when exercising the right to suspend performance for non-payment.

4.12 Off-site materials and goods	4.2.1(c)	Clause redrafted: the clause has been significantly reworded, but does not give rise to new obligations.	If anything, the obligations in this clause are less prescriptive under IC 2005 in that the methods of identifying ownership are not specified, whereas IFC 1998 required various permutations of numbering or letting. Also, the requirement for a bond in respect of goods is stated to be in respect of 'Listed Items which are not uniquely identified' (clause 4.12.5).
4.13 Adjustment of Contract Sum	4.5	Clause redrafted: the clause has been significantly reworded and gives additional clarity but does not give rise to new obligations.	IC 2005 has adopted a slightly more useful layout for clause 4.13.2 than the second paragraph of 4.5 of IFC 1998 but the effect is not changed.
4.14 Issue of Final Certificate	4.6.1–4.6.2	Slight amendment: the clause repeats the text of IFC 1998 with slight changes that improve the drafting or syntax, but the effect is unchanged.	The clause accommodates completion in sections and although it adopts the new defined term 'Final Certificate', the time-scales and processes are unchanged.
Fluctuations			
4.15 Contribution, levy and tax fluctuations	4.9	**Significant change:** the clause has been redrafted and this affects its operation.	The IC 2005 applies the Fluctuations Option set out in detail in Schedule 4 of the contract, unless the Parties have indicated that the Fluctuations Option is not to apply by stating this in the Contract Particulars.
4.16 Fluctuations – Named Sub-Contractors	4.10	Slight amendment: the clause repeats the text of IFC 1998 with slight changes that improve the drafting or syntax, but the effect is unchanged.	
Loss and Expense			
4.17 Disturbance of regular progress	4.11	Clause redrafted: the clause has been significantly reworded and gives additional clarity, but does not give rise to new obligations.	The clause has been redrafted but there are two points of distinction. The first is the use of the new defined term 'Relevant Matter' to describe the events giving rise to a claim for loss and expense.

243

IC 2005 clause and reference title	IFC 1998 clause and reference title	Summary of change	Consequence of change and comments	New actions
			Secondly, the provision relating to deferment of possession has been amended. IFC 1998 referred to 'deferment of the Employer giving possession'. IC 2005 refers to 'deferment of giving possession'. Given that both of these clauses relate to the optional right for the Employer to defer possession (and this must be specifically selected in the Contract Particulars), no particular effect of this change is foreseen.	
4.18 Relevant Matters	4.12	Clause redrafted: the clause has been significantly reworded (and gives additional clarity) but does not give rise to new obligations.	The clause has been substantially redrafted and although there are fewer sub-clauses to this clause than its predecessor in the IFC 1998, the clause has at least the same and perhaps a wider application than clause 4.12 of IFC 1998. The principle distinction is the inclusion of a rather wide catch-all clause at 4.18.5 that might almost be headed 'matters that the Contractor is not responsible for'. Although they are not specifically stated in IC 2005, the provisions of IFC 1998 clauses 4.12.1.1, 4.12.1.2, 4.12.3, 4.12.4, 4.12.6, 4.12.9 are within the scope of clause 4.18.5. Clause 4.18.5 does have limits that show that it is intended as a mechanism of last resort. This is primarily found in clause 4.17, which states that the clause is for 'direct loss and/or expense for which [the Contractor] would not be reimbursed by a payment under any other provision in these Conditions'.	

4.19 Reservation of Contractor's rights and remedies	4.11	No change: the clause repeats the text of IFC 1998.	
Section 5: Variations			
General			
5.1 Definitions of Variations	3.6.1, 3.6.2	Slight amendment: the clause repeats the text of IFC 1998 with slight changes that improve the drafting or syntax, but the effect is unchanged.	
5.2 Valuation of Variations and provisional sum work	3.7.1.1, 3.7.1.2 Option B	Clause redrafted: the clause has been significantly reworded, but does not give rise to new obligations.	The clause has been redrafted and whilst the provision that the Parties agree the value appears to have acquired a greater imperative, the clause is no more exacting than clause 3.7.1.1 of IFC 1998. In addition, the statement that an agreement be reached no longer refers to this being 'prior' agreement. The clause also contains the provision from 3.7.1.2 Option B that unless agreed, the Valuation is to be carried out by the Quantity Surveyor.
The Valuation Rules		The mechanism for a Contractor's price statement has been deleted.	
5.3 Measurable Work	3.7.1.2 Option B		
5.3.1	3.7.1.2 generally	Clause redrafted: the clause has been significantly reworded (and gives additional clarity) but does not give rise to new obligations.	The new layout of the clause is an improvement on the previous form of the contract.
5.3.1.1	3.7.4(a)	Clause redrafted: the clause has been significantly reworded, but does not give rise to new obligations.	

IC 2005 clause and reference title	IFC 1998 clause and reference title	Summary of change	Consequence of change and comments	New actions
5.3.1.2	3.7.4	Clause redrafted: the clause has been significantly reworded, but does not give rise to new obligations.	The 'fair allowance' referred to may be an allowance in terms of a reduction of the price, as well as an increase of the price.	
5.3.1.3	3.7.5	Slight amendment: the clause repeats the text of IFC 1998 with slight changes that improve the drafting or syntax, but the effect is unchanged.		
5.3.1.4	3.7.4(b)	Slight amendment: the clause repeats the text of IFC 1998 with slight changes that improve the drafting or syntax, but the effect is unchanged.		
5.3.1.5	3.7.4(b)	Slight amendment: the clause repeats the text of IFC 1998 with slight changes that improve the drafting or syntax, but the effect is unchanged.		
5.3.2	5.7.3	Slight amendment: the clause repeats the text of IFC 1998 with slight changes that improve the drafting or syntax, but the effect is unchanged.		
5.4 Daywork	3.7.6	**Major change:** the amendment changes a fundamental process or procedure of the contract compared to IFC 1998.	From the point of view of the Employer (or more particularly his Quantity Surveyor), this clause has been improved compared to IFC 1998. IC 2005's updated version of the clause has greater clarity in expressing when, and the extent to which, daywork rates should apply. Under IC 2005 daywork rates should only be used where the 'Valuation relates to the execution of additional or substituted	

			work which cannot properly be valued by measurement'. Note that daywork rates are to be applied to this extent only. In addition, IC 2005 provides that evidence for the labour must be provided by the Contractor in a timely manner. IFC 1998 merely provided that daywork rates be applied 'Where … appropriate' and did not provide for the submission of any voucher-based evidence.	
5.5 Change of conditions for other work	3.7.9	Clause redrafted: the clause has been significantly reworded but does not give rise to new obligations.		
5.6 Additional provisions	3.7.5, 3.7.8	Slight amendment: the clause repeats the text of IFC 1998 with slight changes that improve the drafting or syntax, but the effect is unchanged.		
Section 6: Injury, Damage and Insurance				
Injury to Persons and Property				
6.1 Liability of Contractor – personal injury or dcath	6.1.1	Clause redrafted: the clause has been significantly reworded and gives additional clarity, but does not give rise to new obligations.	The clause has been moderately redrafted – it makes use of the new definition 'Employer's Persons'. More significantly, the IC 2005 version of this clause removes the statement that the liability indemnified is 'under statute or common law' – in so doing, it removes any restriction that might conceivably have followed from such a statement.	
6.2 Liability of Contractor – injury or damage to property	6.1.2	Slight amendment: the clause repeats the text of IFC 1998 with slight changes that improve the drafting or syntax, but the effect is unchanged.	Save for the use of the new term 'Contractor's Person', the clause is unaltered.	

JCT 2005 Intermediate Building
Contract (IC 2005)

IC 2005 clause and reference title	IFC 1998 clause and reference title	Summary of change	Consequence of change and comments	New actions
6.3 Injury or damage to property – Works and Site Materials excluded	6.1.3, 6.1.4	Slight amendment: the text of IFC 1998 is repeated, new defined terms are used and the clause now accommodates the incorporation of the Sectional Completion provisions; notwithstanding this, the effect of the clause is unchanged.		
Insurance against Personal Injury and Property Damage				
6.4 Contractor's insurance of his liability	6.2.1, 6.2.2, 6.2.3	**Significant change:** the clause repeats most of the text of the IFC 1998, but it is also amended in a way that does affect the operation of the clause.	The parallel clauses in IFC 1998 made specific reference to persons with contracts of service and apprenticeships, whereas this version refers to employees only. In response to this, Parties may wish to consider whether their apprentices are employees and the status of self-employed staff.	Confirm that apprenticed staff are covered by insurance and consider whether self-employed individuals are covered.
6.5 Contractor's insurance of liability of Employer	6.2.4	**Major change:** the amendment changes a fundamental process or procedure of the contract compared to IFC 1998.	Note that clause 6.5.2 of IC 2005 requires that the details of the insurance are to be lodged with the Employer, whereas the comparable clause in IFC 1998 required these documents to be lodged with the Contract Administrator.	Contractor to deposit insurance policy and premium receipts with the Employer.
6.6 Excepted Risks	6.2.5	Slight amendment: the clause repeats the text of IFC 1998 with slight changes that improve the drafting or syntax, but the effect is unchanged.		

Insurance of the Works

6.7 Insurance Options	6.3.1	Slight amendment: the clause repeats the text of IFC 1998 with slight changes that improve the drafting or syntax, but the effect is unchanged.		
6.8 Related definitions	6.3.2, 8.3	**Major change:** the amendment changes a fundamental process or procedure of the contract compared to IFC 1998.	The vast majority of the clause is unchanged from the IFC 1998 – the changes are limited to those set out below. The most significant change is in the definition of 'Terrorism Cover'. The change is that the mechanism of loss by terrorism has been de-specified – that is to say, the loss is no longer limited to that caused by 'fire or explosion caused by terrorism' as it was under IFC 1998, thus greatly widening the range of events that ought to be covered by insurance. Other than this, the definitions are straightforward; there are a small number of changes to the terms, but the effect is minimal. Some widen the scope (e.g. the mechanism for the escape of water is no longer defined in the definition of 'Specified Perils', which is sensible); others reflect an update in the language used (e.g. 'tempest' is no longer one of the Specified Perils, presumably on the grounds that 'storm' is now thought adequately to describe the same Specified Peril).	Where the insurance market is 'difficult' Contractors may struggle to obtain the expansive cover indicated by IC 2005 at economic rates, or indeed at all. Where cover proves difficult to obtain, Contractors should follow the procedures in this clause and indicate to Employers such cover as can be obtained. The Parties might then seek agreement as to the available cover, as the alternative is to run the risk of acting in breach of contract. See also clause 6.10.
6.9 Sub-contractors – Specified Periods cover under Joint Names All Risks Policies	6.3.3	Clause redrafted: the clause has been significantly reworded and gives additional clarity but does not give rise to new obligations.		

IC 2005 clause and reference title	IFC 1998 clause and reference title	Summary of change	Consequence of change and comments	New actions
6.10 Terrorism Cover – non-availability – Employer's options	6.3A.5.1 – 6.3A.5.3	Clause redrafted: the clause has been significantly reworded and gives additional clarity but does not give rise to new obligations.	The redraft deals with the Terrorism Cover being provided by either Party more efficiently than this matter was dealt with in IFC 1998. The clause has updated some of its definitions, for example the term 'Effective Date' stated in IFC 1998 has been replaced by 'cessation date' to describe when the Terrorism Cover lapses. In addition, the defined term 'Insurers' Notification' is not used.	
Joint Fire Code – Compliance				
6.11 Application of clause	6.3FC.1	Slight amendment: the clause repeats the text of IFC 1998 with slight changes that improve the drafting or syntax, but the effect is unchanged.		
6.12 Compliance with Joint Fire Code	6.3FC.2.1, 6.3FC.2.2	Clause redrafted: the clause has been significantly reworded and gives additional clarity, but does not give rise to new obligations.	The effect of the clause is unchanged save for the use of the new defined terms 'Employer's Person' and 'Contractor's Person'.	
6.13 Breach of Joint Fire Code – Remedial Measures	6.3FC.3.1	**Significant change:** the clause repeats most of the text of the IFC 1998, but it is also amended in a way that does affect the operation of the clause.	IFC 1998 provided for a 'Remedial Measures Completion Date'. For IC 2005 this provision has been replaced by the recognition in the clause that the requirements stated by the insurer will prevail. To this extent, the clause is less objective but far more pragmatic in terms of the actual processes that apply in such situations. In addition, the recoverability of cost by the Employer in the event of the Contractor's failure to remedy the Works	

			has been refined and the Employer is limited to recovering 'additional cost' by IC 2005. IFC 1999 was less restrictive and provided for the recovery of 'all costs incurred in connection with such employment'.	
6.14 Joint Fire Code – amendments/revisions	6.3FC.5	Slight amendment: the clause repeats the text of IFC 1998 with slight changes that improve the drafting or syntax, but the effect is unchanged.		
Section 7: Assignment and Collateral Warranties				
7.1 Assignment	3.1	Slight amendment: the clause repeats the text of IFC 1998 with slight changes that improve the drafting or syntax, but the effect is unchanged.	Like IFC 1998, IC 2005 provides no free right of assignment – any assignment must have the consent of the other Party. Note also that assignment without consent is also a termination event.	
Collateral Warranties				
7.2 References	N/A	**New definition:** this clause provides a new definition of a term or of a process.		Complete Part 2 of the Contract Particulars with clarity and define the classes of person with precision if collateral warranties are required.
7.3 Notices	N/A	**Major change:** the amendment changes a fundamental process or procedure of the contract compared to IFC 1998.	The requirements for giving proper notice under this clause are stated here. It is suggested that if a JCT Collateral Warranty were to be amended, then the warranty would not be a 'specified JCT Collateral Warranty' and the amended form would have to accompany the notice.	

IC 2005 clause and reference title	IFC 1998 clause and reference title	Summary of change	Consequence of change and comments	New actions
7.4 Execution of Collateral Warranties	N/A	**Major change:** the amendment changes a fundamental process or procedure of the contract compared to IFC 1998.	This clause ensures that the collateral warranties required follow the main contract so that a warranty executed by deed is not required where the main contract is a simple contract , and vice versa. Most readers will be aware that the limitation period for an action for a breach of contract for a contract executed as a simple contract is six years and that the period is 12 years for a deed.	
7.5 Contractor's Warranties – Purchasers and Tenants	N/A	**Major change:** the amendment changes a fundamental process or procedure of the contract compared to IFC 1998.	The Contract Particulars must specifically state this requirement if it is to apply in the absence of any statement there is no obligation to provide a warranty. The notice must comply with clause 7.3 by being in writing and given by actual delivery (i.e. delivered by hand rather than mailed), or by special or recorded delivery. Ordinary post or fax does not suffice. The notice must also identity the person acquiring the rights and their interest is to be stated, that is to say the nature of their proximity or relevance to the Works.	If Purchaser and Tenant rights are required, then this requirement is to be stated in Part 2 of the Contract Particulars before the contract is executed.
7.6 Contractor's Warranty – Funder	N/A	**Major change:** the amendment changes a fundamental process or procedure of the contract compared to IFC 1998.	The Contract Particulars must state this requirement if such a warranty is required – the default position is that no warranty is required. The rights vest at the time the Employer gives the notice to the Contractor. The notice must comply with clause 7.3 by being in writing and given by actual delivery (i.e. delivered by hand rather than mailed), or by special or recorded delivery. Ordinary post or fax does not suffice.	Consider whether a Collateral Warranty is required – if it is required, then state the requirement in Part 2 of the Contract Particulars.

| 7.7 Sub-Contractors' Warranties – Purchasers and Tenants/Funder | N/A | **Major change:** the amendment changes a fundamental process or procedure of the contract compared to IFC 1998. | The Contract Particulars must state that collateral warranties are required from sub-contractors in favour of the Purchasers and Tenants/Funder, otherwise there is no obligation for the provision of such warranties.
The obligation to provide the collateral warranty takes effect upon the receipt of a notice that complies with clause 7.3 by being in writing and given by actual delivery (i.e. delivered by hand rather than mailed), or by special or recorded delivery. Ordinary post or fax does not suffice. The notice also has to state which sub-contractor the collateral warranty is required from, in whose favour and in which form.
The notice must also enclose the terms of the collateral warranty if it is different from, or has amendments to, the JCT Collateral Warranty (clause 7.3).
The Contractor must achieve the execution of the Warranty by its sub-contractor within 21 days of the Employer's notice. The clause specifically considers that the warranty may be provided in different terms. The Employer's approval of such amendments is required, however he cannot unreasonably delay or withhold the consent.
The result is that small-scale changes would ultimately have to be accepted by the Employer. If the Employer states specific requirements in the Contract Particulars, then it is suggested that it would not be unreasonable to withhold consent if the warranty proposed does not satisfy the requirements. | If the Employer requires Sub-Contractor collateral warranties, then this requirement is to be stated in Part 2 of the Contract Particulars before the contract is executed. |

IC 2005 clause and reference title	IFC 1998 clause and reference title	Summary of change	Consequence of change and comments	New actions
7.8 Sub-Contractors' Warranties – Employer		**Major change:** the amendment changes a fundamental process or procedure of the contract compared to IFC 1998.	The Contract Particulars must state that collateral warranties are required from sub-contractors in favour of the Employer. If the Contract Particulars have not been amended to this effect, there will be no obligation to give any collateral warranties. The obligation to provide the warranty takes effect upon the receipt of a notice that complies with clause 7.3 by being in writing and given by actual delivery (i.e. delivered by hand rather than mailed), or by special or recorded delivery. Ordinary post or fax does not suffice. The notice also has to state which sub-contractor the collateral warranty is required from, in whose favour and in which form. The notice must also enclose the terms of the collateral warranty if it is different from or has amendments to the JCT Collateral Warranty (clause 7.3). The Contractor must achieve the execution of the Warranty by its sub-contractor within 21 days of the Employer's notice. The clause specifically considers that the collateral warranty may be provided in different terms. Although the Employer's approval of the amendments is required, he cannot unreasonably delay or withhold the consent. The result is that small-scale changes would ultimately have to be accepted by the Employer. If the Employer states specific requirements in the Contract Particulars then it is suggested that it would not be unreasonable for the Employer to withhold consent if the collateral warranty proposed does not satisfy the requirements.	If the Employer requires sub-contractor collateral warranties, then this requirement is to be stated in the Part 2 of the Contract Particulars before the contract is executed.

Section 8: Termination				
General				
8.1 Meaning of Insolvency	7.3.1	**Major change:** the amendment changes a fundamental process or procedure of the contract compared to IFC 1998.	The definition of 'Insolvency' has been updated and now covers non-incorporated Parties more effectively and appears to extend the definition to cover some more of the preparatory steps to insolvency.	
8.2 Notices under section 8	7.1, 7.2.4, 7.9.5, 7.13.3	Clause redrafted: the clause has been significantly reworded, but does not give rise to new obligations.		
8.3 Other rights, reinstatement	7.8, 7.12	Slight amendment: the clause repeats the text of IFC 1998 with slight changes that improve the drafting or syntax, but the effect is unchanged.	Clause 8.3.2 states that the Parties may agree to a reinstatement of the Contractor. To some extent this is an enabling clause, but ultimately there is no obligation even to consider a reinstatement and even if there were no such clause, a reinstatement on terms would be available to the Parties if they could agree.	
Termination by Employer				
8.4 Default by Contractor	7.2.1	Slight amendment: the clause repeats the text of IFC 1998 with slight changes that improve the drafting or syntax, but the effect is unchanged.		
8.5 Insolvency of Contractor	7.3.1, 7.3.2,	**Major change:** the amendment changes a fundamental process or procedure of the contract compared to IFC 1998.	If the Contractor is Insolvent, the Employer can terminate immediately by notice. The onus is therefore upon the Employer to act swiftly and decisively in the instance of the insolvency of the Contractor.	Notice must be given to the Contractor by the Employer if he wishes to terminate the Contract in the event of the Contractor's Insolvency.

IC 2005 clause and reference title	IFC 1998 clause and reference title	Summary of change	Consequence of change and comments	New actions
			This is a major change compared to IFC 1998 where determination occurred immediately and without any action being required of the Employer in almost all cases. In the event of an Insolvency of a Party to IC 2005, then irrespective of whether notice is given, the contract does operate to suspend the Employer's obligation to pay the Contractor.	The Contractor is obliged to tell the Employer if he is approaching insolvency.
8.6 Corruption	7.4	Clause redrafted: the clause has been significantly reworded, but does not give rise to new obligations.		
8.7 Consequences of termination under clauses 8.4 to 8.6	7.6(a), 7.6(b), 7.6(g)	Clause redrafted: the clause has been significantly reworded, but does not give rise to new obligations.	Clause 8.3.1 has been changed in comparison to clause 7.6(g) of IC 1998. The new clause seeks to cover the same matter but is written in less restrictive terms and provides that the Employer is permitted to recover against 'direct loss and/or damage caused to the Employer (and) for which the Contractor is liable, whether arising as a result of the termination or otherwise'.	
8.8 Employer's decision not to complete the Works	7.7.1	**Significant change:** the clause repeats most of the text of the IFC 1998, but it is also amended in a way that does affect the operation of the clause.	In addition, the redrafting of clause 7.7.1(a) and (b) the function has been amended to oblige the Employer to issue a statement of account if it does not progress the Works within six months of termination. This has made clause 7.7.2 of IFC 1998 unnecessary and thus the mechanism for the Contractor to prompt and ultimately require the Employer to account has been removed.	The Employer is now obliged to account if it does not complete the Works.

Termination by Contractor				
8.9 Default by Employer	7.9.2. 7.9.3, 7.9.4	**Major change:** the amendment changes a fundamental process or procedure of the contract compared to IFC 1998.	Although the clause has been generally redrafted, the grounds of termination for the default of the Employer are unchanged. The change relates to the period of suspension (clause 8.9.2). IFC 1998 provided for a right of termination upon the expiry of a period of one month, whereas IC 2005 requires the Parties to state their own period in the Contract Particulars, but provides a two-month default in the event that they fail or opt not to state a period. The clauses dealing with delay caused by the 'Employer's Persons' is a more economic way of drafting the provision and has made the clause shorter. There is no change to the time which the Employer has to act on a notice nor the ability for the Contractor to act on a default or suspension previously notified.	
8.10 Insolvency of Employer	7.10.1, 7.10.2, 7.10.3	Clause redrafted: the clause has been significantly reworded, but does not give rise to new obligations.		
8.11 Termination by either Party	7.13.1, 17.13.2	**Major change:** the amendment changes a fundamental process or procedure of the contract compared to IFC 1998.	As a consequence of the redrafting the clause has gained brevity. The only significant change relates to the period of suspension (clause 7.13.1). IFC 1998 provided for a right of termination upon the expiry of a period of three months, whereas IC 2005 requires the Parties to state their own period in the Contract Particulars but provides a two-month default in the event that they fail or opt not to state a period. It should be noted that the termination right as a result of 'hostilities' (clause 7.13.1(e)) has been omitted.	

IC 2005 clause and reference title	IFC 1998 clause and reference title	Summary of change	Consequence of change and comments	New actions
8.12 Consequences of Termination under clauses 8.9 to 8.11	7.14, 7.18	**Major change:** the amendment changes a fundamental process or procedure of the contract compared to IFC 1998.	IC 2005 places an obligation upon the Contractor to draw up an account in the event of termination of the contract. The exception to this rule is found in clause 8.7.3 where, following the Employer's termination of the contract for reason of the Contractor's default, the Employer is compelled to draw up the account. In all other instances the responsibility to compile the account lies with the Contractor – this is a change from IFC 1998 where the responsibility usually went to the Party terminating the Agreement. There is no provision for the Employer to withhold an element of retention against the sums that were to be paid against the account.	
Section 9: Settlement of Disputes				
9.1 Mediation	Reference to mediation was previously contained in a footnote	**Major change:** this is a wholly new clause for the IC 2005.	The clause is a statement that the Parties may have their differences resolved by mediation – it is not an obligation that they must do so, nor that they must even consider doing so. This statement in a clause of the contract is new for IC 2005 and also reflects a trend supported by the courts.	
9.2 Adjudication	9A	**Major change:** the amendment changes a fundamental process or procedure of the contract compared to IFC 1998.	The adjudication procedure set out at length in IFC 1998 has been deleted. The IC 2005 adopts the Scheme set out in the *Scheme for Construction Contracts (England and Wales) Regulations* 1998 ('the Scheme').	Users should familiarise themselves with the Scheme for Construction Contracts before contemplating an adjudication process.

		The operation of the Scheme is slightly amended. Firstly by the provision in IC 2005 that the Parties may identify their own Adjudicator or Adjudicator Nominating Body if the necessary amendment is made in the Contract Particulars (this is the same as IFC 1998). Secondly, clause 9.2.2 also departs from the Scheme and provides that an Adjudicator of a dispute concerning clause 3.15 and the reasonableness of instructions under it must either have specialist knowledge or must appoint a specialist advisor. In addition to the removal of the JCT's adjudication process, it is no longer a condition that the Adjudicator has to accept the JCT Adjudicators Agreement that provided terms between the Parties and the Adjudicator. All references to it have been removed. This is a positive change that has removed a potential difficulty in the Adjudicator selection process. Problems occasionally arose if Adjudicators were not willing to use the JCT Adjudication Agreement.	
Arbitration		*Generally:* Note that the resolution of disputes by Arbitration is not the default position under IC 2005 and unless the Parties make the necessary changes in the Contract Particulars these clauses will not apply. The Parties to a contract that does not adopt arbitration may still have a dispute arbitrated if they both agree.	A decision is required prior to the formation of the contract as to whether arbitration or litigation is the default dispute resolution process required. The relative merits of each method should be considered.

IC 2005 clause and reference title	IFC 1998 clause and reference title	Summary of change	Consequence of change and comments	New actions
			In such a circumstance, the Parties would not be arbitrating 'pursuant to Article 8'; thus, unless they also agreed to accept them, these clauses would not apply and the Parties would also have to agree the appropriate arbitration rules.	
9.3 Conduct of arbitration	9B, 9B.6	Slight amendment: the clause repeats the text of IFC 1998 with slight changes that improve the drafting or syntax, but the effect is unchanged.	The clause is updated to incorporate the latest version of the Construction Industry Model Arbitration Rules (CIMAR) rules and merges the provisions of the two previous clauses from IFC 1998, but is otherwise the same.	
9.4 Notice of reference to arbitration	9B.1.1, 9B.1.2(a) and (b)	Clause redrafted: the clause has been significantly reworded, but does not give rise to new obligations.	The method and process for giving notice of Arbitration is stated here. The distinction is that IFC 1998 reproduced the relevant part of the CIMAR (e.g. Rules 2.1, 2.3 and 2.5) whereas IC 2005 does not.	
9.5 Powers of Arbitrator	9B.2	Clause redrafted: the clause has been significantly reworded, but does not give rise to new obligations.		
9.6 Effect of award	9B.3	Clause redrafted: the clause has been significantly reworded, but does not give rise to new obligations.		
9.7 Appeal – questions of law	9B.4	Clause redrafted: the clause has been significantly reworded, but does not give rise to new obligations.		
9.8 Arbitration Act 1996	9B.5	Clause redrafted: the clause has been significantly reworded, but does not give rise to new obligations.		

SCHEDULES

Schedule 1: Insurance Options			
Insurance Option: New Buildings – All Risks Insurance of the Works by the Contractor			
A.1 Contractor to take out and maintain a Joint Names Policy	6.3A.1	Clause redrafted: the clause has been significantly reworded and accommodates the incorporation of the Sectional Completion provisions, but does not give rise to new obligations.	In addition to repeating the text of clause 6.3A.1 of IFC 1998, the clause also contains the requirement for the insurer to be to the Employer's approval from clause 6.3A.2. Note that the second paragraph of IFC 1998 clause 6.3B.1 has been deleted. This clause was largely explanatory, or even descriptive, as it dealt with the mechanism by which a VAT-exempt Employer would manage VAT payments in the event of damage to the Works. The explanation of this process is now located in the guidance note. Note also that an additional paragraph has been added to accommodate the sectional completion of the Works. The consequence of the change is that the Contractor is not obliged to insure Sections that have been handed over.
A.2 Insurance documents – failure by Contractor to insure	6.3A.2	Slight amendment: the clause repeats the text of IFC 1998 with slight changes that improve the drafting or syntax, but the effect is unchanged.	
A.3 Use of Contractor's annual policy – as alternative	6.3A.3.1	Slight amendment: the clause repeats the text of IFC 1998 with slight changes that improve the drafting or syntax, but the effect is unchanged.	

261

IC 2005 clause and reference title	IFC 1998 clause and reference title	Summary of change	Consequence of change and comments	New actions
A.4 Loss or damage, insurance claims and Contractor's obligations	6.3A.4.1	Slight amendment: the clause repeats the text of IFC 1998 with slight changes that improve the drafting or syntax, but the effect is unchanged.		
A.5 Terrorism Cover – premium rate changes	6.3A.3.1	**Significant change:** the clause repeats most of the text of the IFC 1998, but it is also amended in a way that does affect the operation of the clause.	The change to the clause is found in A.5.2 where the mechanism of loss by terrorism is stated. The IC 2005 version of the clause is a great deal wider than the previous version of the contract, which limited the mechanism to 'physical loss or damage caused by fire or explosion caused by terrorism'. The present definition is de-specified and is not limited to loss caused by means of explosion or fire.	
Insurance Option B: New Buildings – All Risks Insurance of the Works by the Employer				
B.1 Employer to take out and maintain a Joint Names Policy	6.3B.1	Clause redrafted: the clause has been significantly reworded and accommodates the incorporation of the Sectional Completion provisions, but does not give rise to new obligations.	In addition to repeating the text of clause 6.3B.1 of IFC 1998 the clause has two significant points to be noted. The second paragraph of IFC 1998 clause 6.3B.1 has been deleted. This clause was largely explanatory, or even descriptive, as it dealt with the mechanism by which a VAT-exempt Employer would manage VAT payments in the event of damage to the Works. The explanation of this process is now located in the guidance note. Note also that an additional paragraph has been added to accommodate the sectional completion of the Works. The	

			consequence of the change is that the Employer is not obliged to maintain the insurance (at least as a Joint Names Policy for the purposes of this Agreement) for Sections that have been handed over.	
B.2 Evidence of Insurance	6.3B.2, 6.3B.4.4	Slight amendment: the clause repeats the text of IFC 1998 with slight changes that improve the drafting or syntax, but the effect is unchanged.	Note that the two, previously separate clauses, have been joined together.	
B.3 Loss or damage, insurance claims, Contractor's obligations and payment by Employer	6.3B.3.1	Slight amendment: the clause repeats the text of IFC 1998 with slight changes that improve the drafting or syntax, but the effect is unchanged.	In this instance clause B3.2 has been clarified to make it clear the rectification work done by the Contractor after an insured event is to be paid for as a Variation.	
Insurance Option C: Insurance by the Employer of Existing Structures and Works in or Extensions to them				
C.1 Existing structures and contents – Joint Names Policy for Specified Perils	6.3C.1	Clause redrafted: the clause has been significantly reworded and accommodates the incorporation of the Sectional Completion provisions, but does not give rise to new obligations.	In addition to repeating the text of clause 6.3C.1 of IFC 1998, the clause has two significant points to be noted. The second paragraph of IFC 1998 clause 6.3C.1 has been deleted. This clause was largely explanatory, or even descriptive, as it dealt with the mechanism by which a VAT-exempt Employer would manage VAT payments in the event of damage to the Works. The explanation of this process is now located in the guidance note. Note also that an additional paragraph has been added to accommodate the sectional completion of the Works. The consequence of the change is that the Employer is not obliged to maintain the insurance (at least as a Joint Names Policy for the purposes of this Agreement) for Sections that have been handed over.	

263

IC 2005 clause and reference title	IFC 1998 clause and reference title	Summary of change	Consequence of change and comments	New actions
C.2 The Works – Joint Names Policy for All Risks	6.3C.2	Slight amendment: the clause repeats the text of IFC 1998 with slight changes that improve the drafting or syntax, but the effect is unchanged.		
C.3 Evidence of Insurance	6.3C.3, 6.3C.5.4	Slight amendment: the clause repeats the text of IFC 1998 with slight changes that improve the drafting or syntax, but the effect is unchanged.		
C4 Loss or damage to Works – insurance claims and Contractor's obligations	6.3C.4.1	Clause redrafted: the clause has been significantly reworded and gives additional clarity but does not give rise to new obligations.		
Schedule 2: Named Sub-Contractors				
1	3.3.1	Slight amendment: the clause repeats the text of IFC 1998 with slight changes that improve the drafting or syntax, but the effect is unchanged.	The first sentence of the last paragraph of clause 3.3.1 of IFC 2005 is repeated here.	
2	3.3.1	Slight amendment: the clause repeats the text of IFC 1998 with slight changes that improve the drafting or syntax, but the effect is unchanged.		
3	3.3.1	Slight amendment: the clause repeats the text of IFC 1998 with slight changes that improve the drafting or syntax, but the effect is unchanged.		
4	3.3.1	Slight amendment: the clause repeats the text of IFC 1998 with slight changes that improve the drafting or syntax, but the effect is unchanged.		

5	3.3.2	Slight amendment: the clause repeats the text of IFC 1998 with slight changes that improve the drafting or syntax, but the effect is unchanged.		
6	3.3.3	Slight amendment: the clause repeats the text of IFC 1998 with slight changes that improve the drafting or syntax, but the effect is unchanged.		
7	3.3.3	Slight amendment: the clause repeats the text of IFC 1998 with slight changes that improve the drafting or syntax, but the effect is unchanged.		
8	3.3.4	Clause redrafted: the clause has been significantly reworded but does not give rise to new obligations.		
9	3.3.5	Slight amendment: the clause repeats the text of IFC 1998 with slight changes that improve the drafting or syntax, but the effect is unchanged.		
10	3.3.6	Slight amendment: the clause repeats the text of IFC 1998 with slight changes that improve the drafting or syntax, but the effect is unchanged.		
11	3.3.7	Slight amendment: the clause repeats the text of IFC 1998 with slight changes that improve the drafting or syntax, but the effect is unchanged.		
12	3.3.8	Clause redrafted: the clause has been significantly reworded but does not give rise to new obligations.		
13	3.3.9	Clause redrafted: the clause has been significantly reworded but does not give rise to new obligations.		

IC 2005 clause and reference title	IFC 1998 clause and reference title	Summary of change	Consequence of change and comments	New actions
Schedule 3: Forms of Bonds				
Part 1: Advance Payment Bond	Annex 1 to Appendix: Terms of Bonds: Advance Payment Bond	No change: the clause repeats the text of IFC 1998, but note that the execution provisions have been adjusted.	There is a minor deviation in the execution provisions of the Advance Payment Bond in that the Bond is now executed by a person acting 'as Attorney and on behalf of the Surety' rather than a person acting 'for and on behalf of the Surety'. In addition, individuals witnessing the execution must state their name and address. Other than these formalities, the terms of the Bond and 'Notice of Demand' are identical.	Surety's representative must now be sufficiently empowered to act as Attorney of the Surety. Ensure that witnesses confirm their identity and provide an address.
Part 2: Bond in respect of payment for off-site materials and/or goods	Annex 1 to Appendix: Terms of Bonds: Bond in respect of payment for off-site materials and/or goods	Slight amendment: the clause repeats the text of IFC 1998 with slight changes that improve the drafting or syntax, but the effect is unchanged. Note that the execution provisions have been adjusted.	A new clause 11 has been added to exclude the provisions of the *Contracts (Rights of Third Parties) Act* 1998 – given that the original bond was not intended to provide rights for third parties, this amendment improves its application. In addition, clause 3.1 has been slightly adjusted. There is also a minor deviation in the execution provisions of the Bond in that it is now to be executed by a person acting as 'Attorney and on behalf of the Surety' rather than a person acting 'for and on behalf of the Surety'. In addition, individuals witnessing the execution must state their name and address. Other than these formalities, the terms of the Bond and Notice of Demand are unchanged.	Surety's representative must now be sufficiently empowered to act as Attorney of the Surety. Ensure that witnesses confirm their identity and provide an address.

Schedule 4: Fluctuations Option – Contribution, levy and tax fluctuations			*Generally:* IFC 1998 did not include the provisions of the Fluctuation clauses – a separate supplement had to be purchased. In order to determine what is new, IC 2005 has been compared with Intermediate Form of Building Contract 1998 Edition Incorporating Amendments 1 to 3: Fluctuation clauses, Supplemental Condition C.	
1 Deemed calculation of Contract Sum – labour	C1	Slight amendment: the clause repeats the text of IFC 1998 Supplemental Condition C but makes use of new defined terms and makes slight changes, but the effect is unchanged.	References to legislation have been updated. Thus IC 2005 refers to the *Industrial Training Act* of 1982 rather than the act of the same name of 1962. Similarly, reference is now made to 1993 *PAYE Regulations* (rather than the *PAYE Regulations* of 1973) and the *Pensions Schemes Act* of 1993 (instead of the *Social Security Act* of 1975).	
2 Deemed calculation of Contract Sum – materials	C2	Slight amendment: the clause repeats the text of IFC 1998 Supplemental Condition C but makes use of new defined terms and makes slight changes, but the effect is unchanged.		
3 Sub-let work – incorporation of provisions to like effect	C3	No change: the clause repeats the text of IFC 1998 Supplemental Condition C but makes use of new defined terms		
4 Written notice by Contractor	C4	Slight amendment: the clause repeats the text of IFC 1998 Supplemental Condition C but makes use of new defined terms and makes slight changes, but the effect is unchanged.	Note a minor typographical error has been corrected in the new text.	
5 Agreement – Quantity Surveyor and Contractor	C4.3	No change: the clause repeats the text of IFC 1998 Supplemental Condition C.		

IC 2005 clause and reference title	IFC 1998 clause and reference title	Summary of change	Consequence of change and comments	New actions
6 Fluctuations added to or deducted from Contract Sum	C4.4	Slight amendment: the clause repeats the text of IFC 1998 Supplemental Condition C but makes use of new defined terms and makes slight changes, but the effect is unchanged.		
7 Evidence and computation by Contractor	C4.5	No change: the clause repeats the text of IFC 1998 Supplemental Condition C.		
8 No alteration to Contractor's profit	C4.6	No change: the clause repeats the text of IFC 1998 Supplemental Condition C.		
9 Position where Contractor in default over completion	C4.7	Slight amendment: the clause repeats the text of IFC 1998 Supplemental Condition C but makes use of new defined terms and makes slight changes, but the effect is unchanged.		
10 Work etc to which paragraphs 1 to 3 not applicable	C5	Slight amendment: the clause repeats the text of IFC 1998 Supplemental Condition C but makes use of new defined terms and makes slight changes, but the effect is unchanged.	Note that the IC 2005 now explains that Contractors must apply the terms of their own contracts with their sub-contractors where relevant.	
11 Definitions	C6	No change: the clause repeats the text of IFC 1998 Supplemental Condition C.		
12 Percentage addition to fluctuation payments or allowances	C7	No change: the clause repeats the text of IFC 1998 Supplemental Condition C.		

Contract clause headings and numbering structure from the *Intermediate Building Contract*, Joint Contracts Tribunal Limited, 2005, Sweet and Maxwell, © The Joint Contracts Tribunal Limited 2005, are reproduced here with permission.

Table of destinations: IFC 1998 and IC 2005

This table of destinations follows the structure of the old IFC 1998 and states the new clause reference in the IC 2005 form.

IFC 1998	IC 2005	IFC 1998	IC 2005
Parties to Agreement	Parties to Agreement	1.6	2.8
First recital	First, Second and Third recitals	1.7.1	2.10
Second recital	Fourth recital	1.7.2	2.11
Third recital	Seventh recital	1.8	2.8
Fourth recital	Sixth recital	1.9	1.9
Fifth recital		1.10	2.17
Article 1	Article 1	1.11	2.18
Article 2	Article 2	1.12	3.19
Article 3	Articles 3 and 3.4	1.13	1.7
Article 4	Articles 4 and 3.4	1.14	1.5
Article 5	Article 5	1.15	1.12
Article 6	Article 6	1.16	1.8
Article 7	Eighth recital	1.17	1.6
Article 8	Article 7	2.1	2.4 and 2.6
Article 9A	Article 8	2.2	2.5
Article 9B	Article 9	2.3	2.19
Attestation	Attestation	2.4	2.20
1.1	2.1 and 2.2	2.5	2.19.5
1.2	2.2 and 4.1	2.6	2.22
1.3	1.3	2.7	2.23
1.4	2.12 and 2.13	2.8	2.24
1.5	2.12	2.9	2.21

IFC 1998	IC 2005
2.10	2.30 and 2.31
2.11	2.25, 2.26, 2.27, 2.28 and 2.29
3.1	7.1
3.2	3.5 and 3.6
3.2.1	3.6
3.2.3	3.6
3.3.1	3.7 and Schedule 2, paras 1, 2, 3 and 4
3.3.2	Schedule 2, para 5
3.3.3	Schedule 2, paras 6 and 7
3.3.4	Schedule 2, para 8
3.3.5	Schedule 2, para 9
3.3.6	Schedule 2, para 10
3.3.7	Schedule 2, para 11
3.3.8	Schedule 2, para 12
3.3.9	Schedule 2, para 13
3.4	3.2
3.5.1	3.8 and 3.9
3.5.2	3.10
3.6	3.11
3.6.1	2.14 and 5.1.1
3.6.2	5.1.2
3.7.1.1	5.2
3.7.1.2	5.2, 5.3 and 5.3.1
3.7.4	5.3.1.2
3.7.4(a)	5.3.1.1
3.7.4(b)	5.3.1.4 and 5.3.1.5

IFC 1998	IC 2005
3.7.5	5.3.1.3 and 5.6
3.7.6	5.4
3.7.8	5.6
3.7.9	5.5
3.8	3.13
3.9	2.9
3.10	3.3
3.11	2.7
3.12	3.14
3.13	3.15
3.14.1	3.16
3.14.2	3.16
3.15	3.12
4A	4.4
4.1	4.2
4.2(a)	4.6, 4.8.1 and 4.8.5
4.2(b)	4.5
4.2(c)	4.6
4.2(d)	4.7
4.2.1	4.7.1
4.2.1(a)	4.7.1
4.2.1(c)	4.7.1 and 4.12
4.2.2	4.7.2 and 4.7.3
4.2.3(a)	4.8.2
4.2.3(b)	4.8.3
4.2.3(c)	4.8.4

IFC 1998	IC 2005
4.3	4.9
4.4	4.10
4.4A	4.11
4.5	4.13
4.6.1	4.14
4.6.2	4.14
4.7	1.10
4.8	1.11
4.9	4.15
4.10	4.16
4.11	4.17 and 4.19
4.12	4.18
5.1	2.1 and 2.3
5.2	2.15.1
5.3	2.15.3 and 3.19
5.4	2.16
5.5	4.3
5.6	Fifth recital and 4.4
5.7	3.18
6.1.1	6.1
6.1.2	6.2
6.1.3	6.3
6.1.4	6.3
6.2.1	6.4
6.2.2	6.4
6.2.3	6.4

IFC 1998	IC 2005
6.2.4	6.5
6.2.5	6.6
6.3.1	6.7
6.3.2	6.8
6.3.3	6.9
6.3A.1	Schedule 1, para A1
6.3A.2	Schedule 1, para A2
6.3A.3.1	Schedule 1, para A3
6.3A.4.1	Schedule 1, para A4
6.3A.5.1	6.10
6.3A.5.2	6.10
6.3A.5.3	6.10
6.3A.5.4	Schedule 1, para A5
6.3B.1	Schedule 1, para B1
6.3B.2	Schedule 1, para B2
6.3B.3.1	Schedule 1, para B3
6.3B.4.4	Schedule 1, para B2
6.3C.1	Schedule 1, para C1
6.3C.2	Schedule 1, para C2
6.3C.3	Schedule 1, para C3
6.3C.4.1	Schedule 1, para C4
6.3C.5.4	Schedule 1, para C3
6.3D	
6.3FC.1	6.11
6.3FC.2.1	6.12
6.3FC.2.2	6.12

JCT 2005 Intermediate Building Contract (IC 2005)

IFC 1998	IC 2005
6.3FC.3.1	6.13
6.3FC.5	6.14
7.1	8.2
7.2.1	8.4
7.2.4	8.2
7.3.1	8.1 and 8.5
7.3.2	8.5
7.4	8.6
7.5	
7.6(a)	8.7
7.6(b)	8.7
7.6(g)	8.7
7.7.1	8.8
7.8	8.3
7.9.2	8.9
7.9.3	8.9
7.9.4	8.9
7.9.5	8.2
7.10.1	8.10
7.10.2	8.10
7.10.3	8.10
7.11	
7.12	8.3
7.13.1	8.11
7.13.2	8.11
7.13.3	8.2

IFC 1998	IC 2005
7.14	8.12
7.15	
7.16	
7.17	
7.18	8.12
7.19	
8.1	1.2
8.2	1.3
8.3	1.1 and 6.8
8.4	
8.5	
9A	9.2
9B	9.3
9B.1.1	9.4
9B.1.2(a)	9.4
9B.1.2(b)	9.4
9B.2	9.5
9B.3	9.6
9B.4	9.7
9B.5	9.8
9C	
Supplemental Conditions	
A	
B	
C	Schedule 4
D	

5 JCT Minor Works Building Contract 2005

Introduction

The JCT Minor Works Building Contract 2005 edition (MW 2005) is the successor to the JCT Agreement for Minor Building Works 1998 edition (MBW 1998). MBW 1998 was one of the JCT's most frequently used contracts – if the uptake of the 2005 form is similar to that of the 1998 form, it will remain the true work horse in the JCT stable of contracts.

Indeed, it is likely that the appeal of this contract will increase, as the JCT has separately published a new form of Minor Works Contract with provisions for Contractor's Design for the first time (this is not a full 'design and build' contract but rather a minor works form 'With Contractor's Designed Portion'). Note that this chapter considers the terms of the standard MW 2005 and does not consider the MW with Contractor's Design 2005.

Purpose and structure of this chapter

The first part of this chapter summarises the main changes made in the new contract and the second part of this chapter seeks to analyse the whole of the new contract, to identify and explain the differences between MBW 1998 and MW 2005. The third and final part of this chapter is a table of destinations that states the clauses of MBW 98 and gives the location of the clauses in MW 2005 where there is one. This is a publication for practitioners familiar with the existing JCT contracts and consequently discussion of clauses that have not been changed in the new version of the contract is beyond the scope of this guide.

The second part of this chapter has been set out in a tabular format; the left-hand column states the number and name of the MW 2005 clause and on its right is the number and name (where there is one) of the MBW 1998 clause. A statement of the change and its nature is found in the third column. The fourth column from the left has, where they are of assistance, specific comments against each clause as they arise and includes a note of the consequences of the differences. On the right-hand side is the final column which states any new actions that the Parties must perform and, where relevant, provides guidance on what this means in practice.

The tabular format has been chosen to aid use as a point of reference when contract documents are being prepared and when the contract is used in practice on site.

Base documents

In this chapter, MW 2005 has been compared to MBW 1998 incorporating Amendments 1: 1999, 2: 2000, 3: 2001, 4: 2002 and 5: 2003.

Summary of the changes in MW 2005

The changes that are found in the MW 2005 are incremental and developmental, rather than fundamental. There is a noticeable improvement in the structure, layout and, in places, the drafting of the contract. There seems to have been an effort to minimise the use of legalistic or archaic language and reduce the reliance on supplemental clauses. This effort has been largely successful and should make the contract far more comprehensible.

Although most of the changes relate to improvement or modernisation of the text, there have been some substantive changes that will certainly affect the way in which the contract is operated in practice.

273

Significant changes

Restructuring and redrafting

The layout of the contract has been improved. It is now set out in a single column of text, like the Standard Building Contact (or even like the old JCT 1998) rather than the two-column format adopted by the MBW 1998 and the IFC 1998.

The clauses themselves are grouped into sections and the clauses are given clear titles and the layout is generally less cluttered. This is more than just a housekeeping exercise: it should result in the document being better understood and applied by its users in practice on the basis that if it is possible to discern how the process is to be correctly applied, then opportunities for procedural failure and associated disputes are reduced.

Positive changes to the drafting and greater economy of expression are perhaps slightly obscured by some changes in the names of various familiar terms – for example, the 'Defects Liability Period' has been rechristened the 'Rectification Period'.

Clarification of ambiguous clauses

In general the changes that have been made make the process and the intent clearer but do not add complexity. For example, the process for certification and payment certificates has not been changed so much as the clauses have been stated with additional clarity. The result is a clause that establishes that either, both or only one certificate may act as a withholding notice, provided it has sufficient detail and is served at the correct point in time.

Changes of substance

Contract Particulars

All standard form contracts require the individual details that tailor the standard form to the particular projects to be stated in or appended to the document that is ultimately to be signed. The MW 2005 follows the format of the other contracts in the JCT's 2005 release by seeking to place most of the contract-specific information at the beginning of the contract in what are called the Contract Particulars. MBW 1998 required much the same information but scattered it through the contract. Whist MBW 1998 was not an especially large document, this did have the occasional result that parts would be left incomplete or optional requirements left un-stated.

The contract has also moved towards stating a default position on many of the options that Parties may select or that reflect regulations to be accommodated by the Parties. For example, the Contract Particulars now provide the default position that the Architect/Contract Administrator is the Planning Supervisor for the purpose of the CDM Regulations unless this default position is deviated from.

Dispute resolution

The dispute resolution processes have been comprehensively revised. The adjudication process has been changed substantially in comparison to MBW 1998. The new form provides that the Parties may identify an Adjudicator Nominating Body (ANB) to propose an adjudicator or may identify an individual adjudicator for the contract in the Contract Particulars. The new form also makes better provision for the situation where the section has not been completed at all or left partially blank– in that circumstance a Party requiring a matter to be adjudicated may select any of the nominating bodies stated in the contract to make a nomination.

The process of adjudication has also been changed as JCT Adjudication Rules have been dropped in favour of the Scheme for Construction Contracts; it is thought that this change reflects experience with, and the authority that now supports, the Scheme rather than any fundamental dissatisfaction with the JCT Adjudication Rules.

MBW 1998 stated that arbitration was the default dispute resolution method for all disputes. The most significant change to the MBW dispute resolution mechanism is that MW 2005 reverses this default position. Unless the Parties specifically select arbitration in the Contract Particulars, then it will not be the default mechanism for disputes under the contract. This means that the Parties will use the courts to assert their rights.

The contract also has a new clause that suggests that the Parties consider the mediation of disputes (a process that has found much favour in the courts).

Termination

The termination clauses have been redrafted and the change runs deeper than the change in terminology which adopts the word 'termination' as a replacement for the MBW phrase 'determination'. The definitions of insolvency have been updated and a right for either Party to terminate has been stated. If the Employer has terminated the contract, then there is a new action for the Employer to prepare an account to determine the sums due, rather than wait for the Contractor to take action.

There is no automatic termination in the event of insolvency as there was is most instances of insolvency under MBW 1998. Instead the MW 2005 requires a notice to be issued by the Party wishing to terminate for insolvency in all cases.

Changes to assist in execution

The Minor Works forms are used on small value projects and frequently Parties are reluctant to seek assistance of external advisors either in drafting the contracts or when projects experience difficulty. The preparation and execution of a contract can be a complicated process, and of all forms of the JCT contact, the Minor Works forms are the ones that least frequently cross the desks of lawyers or even professional in-house legal support staff.

In consequence the JCT's notes that accompany the contract in the text and at the end of the terms and conditions have arguably greater significance under this contract than under the other contracts. The fact that these notes have generally been improved should assist in getting the users to choose the correct mode of execution. The contract also includes notes on the process for execution on pages 8 and 10 of MW 2005. The note on page 8 of MW 2005 does indicate that information on choosing between executing the contract as a deed or as a simple contract is in the Guidance Note but does not point out that it refers to section 17 of the Guidance Note at page 35 of MW 2005 under the heading 'Rights and remedies generally'. The requirement to state the company number of the contracting company should assist in providing an absolute way of identifying the contracting entity – this can be especially difficult when contracting with a Party with a number of operating divisions or subsidiaries at various locations.

JCT 2005 Minor Works Building Contract (MW 2005)

MW 2005 clause, reference or title	MBW 1998 clause, reference or title	Summary of change	Consequence of change and comments	New actions
Inside cover			A note on usage is included in the inside cover of the contract. The statements are new and provide guidance which is both simple and direct.	Even users familiar with the Minor Works form would be well-advised to refresh their memory and check the suitability of MW 2005 for the proposed project.
Articles of Agreement	'This Agreement'	**Significant change:** the clause repeats most of the text of MBW 1998, but this is amended in a way that does affect the operation of the clause.	Stating the company number gives additional clarity as to the identity of the Parties, as there is a risk that the company named is not the company whose address is stated. The company number provides an additional reference point that can easily be checked, for example by searching on the Companies House website or using one of the commercial reporting agencies.	Companies should state their company number.
Recitals			*Generally:* The Recitals are similar to MBW 1998, but the order of Recitals has been slightly changed. Some of the information that was provided in the Recitals is now provided in the Articles or the Contract Particulars.	The reference to the quantity surveyor has been omitted from the Recitals of MW 2005. This means that such quantity surveying function as is required, may be undertaken by the Contractor, the Architect/Contract Administrator or the Employer. In practice, if quantity surveying services are

				required they are likely to be provided by a sub-consultant or employee of one of the named Parties.
First Recital (*'the Works'*)	First Recital	Clause redrafted: the clause has been significantly reworded, but does not give rise to new obligations.	Details concerning the identity of the Architect/Contract Administrator are now found in Article 3 and the description of the Contract Documents has been placed in the Second Recital.	
Second Recital (*'the Contract Drawings'*)	First Recital and Third Recital	Clause redrafted: the clause has been significantly reworded, but does not give rise to new obligations.	The Contract Documents are recited and the requirement to mark the Contract Documents is stated here (MBW 1998 stated this requirement in its Third Recital). MW 2005 also makes provision for the Parties to list the documents as well as, or instead of, numbering them.	
Third Recital (*'Documents supplied by the Contractor'*)	Second Recital	Slight amendment: The clause repeats the text of MBW 1998 but makes use of new defined terms and makes slight changes, but the effect is unchanged.	Note also that MBW 1998 recited the provision of the Contract Sum by the Contractor in Article 2 – this has not been repeated in MW 2005, however such recitation is unnecessary and the provision of the Contract Sum is adequately covered in Article 2.	
Fourth Recital (*'the CDM Regulations'*)	Fifth Recital	Clause redrafted: the clause has been significantly reworded but does not give rise to new obligations.	MW 2005 requires the Parties to state the application of the CDM Regulations in the Contract Particulars (two options are provided) MBW 1998 provided for the Recital of the same two options in its Fifth Recital. There should be no consequences as a result of this change.	Employers who are unfamiliar with JCT contracts may need to seek the advice of the Architect/Contract Administrator or a Planning Supervisor in order to complete this Recital.

JCT 2005 Minor Works Building Contract (MW 2005)

MW 2005 clause, reference or title	MBW 1998 clause, reference or title	Summary of change	Consequence of change and comments	New actions
The Articles			*Generally:* Compared to MBW 1998, the new Articles have been slimmed down and words that are more contemporary and precise have been used.	
Article 1: Contractor's Obligations	Article 1	Clause redrafted: the clause has been significantly reworded but does not give rise to new obligations.	The statement is made that: 'The Contractor shall carry out and complete the Works in accordance with the Contract Documents.' The explicit statement that this obligation is undertaken for a consideration has been removed, but a valid contract would nevertheless be created as the exchange of obligations is apparent and adequately dealt with by subsequent Articles and by the terms of the Contract.	
Article 2: Contract Sum	Article 2	Slight amendment: the clause repeats the text of MBW 1998 with slight changes that improve the drafting or syntax, but the effect is unchanged.		
Article 3: Architect/ Contract Administrator	Article 3	Slight amendment: the clause repeats the text of MBW 1998 with slight changes that improve the drafting or syntax, but the effect is unchanged.		
Article 4: Planning Supervisor	Article 4	Clause redrafted: the clause has been significantly reworded and gives additional clarity, but does not give rise to new obligations.	The Architect/Contract Administrator is to be the Planning Supervisor, unless there is a specific appointment to the contrary. Whilst the process is not changed compared to MBW 1998, the amended clause establishes the default position more effectively and avoids the possibility of there not being a Planning Supervisor.	State the identity of the Planning Supervisor unless the Parties wish this to be the Architect/Contract Administrator.

Article 5: Principal Contractor	Article 5	Slight amendment: The clause repeats the text of MBW 1998 with slight changes that improve the drafting or syntax, but the effect is unchanged.	The CDM Regulations apply mandatory obligations upon the person identified as the 'Principal Contractor'. The default position is that the Contractor shall be the Principal Contractor for the purposes of the CDM Regulations. Unlike MBW 1998, MW 2005 provides a space in the Articles for the name of the Principal Contractor to be inserted if this is to be someone other that the Contractor.	Identify the Principal Contractor here if this is not to be the Contractor.
Article 6: Adjudication	Article 6	**Major change:** the amendment changes a fundamental process or procedure of the contract compared to MBW 1998.	A greatly simplified Article states the application of adjudication to any dispute or difference under the contract. The difference compared to MBW 1998 is that the rules for the adjudication have changed. The provisions of the Scheme replace JCT's own process that had previously been found in Supplemental Condition D of MBW 1998. Irrespective of the change, the Parties may still select a specific adjudicator or Adjudicator Nominating Body by stating this in the Contact Particulars.	
Article 7: Arbitration	Article 7A	**Major change:** the amendment changes a fundamental process or procedure of the contract compared to MBW 1998.	This is the most significant change found in the Articles. The contract does not automatically provide for the arbitration of disputes unless the Parties specifically select the arbitration option in the Contract Particulars. This is a change from the default position of MBW 1998. Under that contract all disputes were to be referred to arbitration in preference to the courts and the courts would halt litigation in order to allow the arbitration process to be carried out. Arbitration can still be adopted as the default dispute resolution process by selecting it in the Contract Particulars and the clauses necessary to support this are	Consider the merits of arbitration – no action is necessary if Parties are content to proceed without arbitration.

279

MW 2005 clause, reference or title	MBW 1998 clause, reference or title	Summary of change	Consequence of change and comments	New actions
			also provided in the contract. If adopted, the process to be applied remains the Construction Industry Model Arbitration Rules (CIMAR). The provision or exclusion of arbitration does not have any effect on a Party's statutory right to adjudication under the *Housing Grants, Construction and Regeneration Act* 1996 ('the Construction Act'). Thus, selecting arbitration does not prevent adjudication, nor prevent an adjudication being run concurrently with an arbitration process.	
Article 8: Legal proceedings	Article 7B	Clause redrafted: the clause has been significantly reworded and gives additional clarity, but does not give rise to new obligations.		
Contract Particulars	Appendix		*Generally:* MW 2005 collects the contract-specific information together and creates a single point of reference. The Contract Particulars must be completed in full by the Parties before the contract is executed. Unlike those JCT 1998 standard forms for larger projects, there was not a specific appendix at the end of MBW 1998. Instead of having an appendix some of the contract-specific information was stated in the Articles of further information was in the relevant clauses themselves and some was gathered in clause 2. Grouping the information here as 'Contract Particulars' should improve use in practice and assist in the preparation of contract documents.	If there is a general action, then it is to ensure that the Contract Particulars are given due attention and completed correctly prior to execution of the contract.

Fourth Recital CDM Regulations	Fifth Recital	Slight amendment: the clause repeats the text of MBW 1998 with slight changes that improve the drafting or syntax, but the effect is unchanged.	This provision requires the selection of an express statement of the scope of the application of the CDM Regulations.	
Article 7: Arbitration	Article 7A	**Major change:** the amendment changes a fundamental process or procedure of the contract compared to MBW 1998.	This provision seeks an express statement on whether or not arbitration will apply as the dispute resolution mechanism for the contract. If no preference is stated, the default position is that arbitration will not apply.	Specifically opt in to arbitration if it is required.
2.2 Date for Commencement of the Works	2.1	Slight amendment: the clause repeats the text of MBW 1998 with slight changes that improve the drafting or syntax, but the effect is unchanged.		As with the previous version of the contract, a commencement date should be inserted.
2.2 Date for completion	2.1	Slight amendment: the clause repeats the text of MBW 1998 with slight changes that improve the drafting or syntax, but the effect is unchanged.		As with the previous version of the contract a completion date should be inserted.
2.8 Liquidated Damages	2.3	Slight amendment: the clause repeats the text of MBW 1998 with slight changes that improve the drafting or syntax, but the effect is unchanged.		
2.10 Rectification Period	2.5	Slight amendment: the clause repeats the text of MBW 1998 with slight changes that improve the drafting or syntax, but the effect is unchanged.	Note that the new term 'Rectification Period' is used to describe what used to be called the 'Defects Liability Period'.	
4.3	4.2.1	Slight amendment: the clause repeats the text of MBW 1998 with slight changes that improve the drafting or syntax, but the effect is unchanged.		
4.5	4.3	Slight amendment: the clause repeats the text of MBW 1998 with slight changes that improve the drafting or syntax, but the effect is unchanged.		

MW 2005 clause, reference or title	MBW 1998 clause, reference or title	Summary of change	Consequence of change and comments	New actions
4.8.1	4.5.1.1	Slight amendment: the clause repeats the text of MBW 1998 with slight changes that improve the drafting or syntax, but the effect is unchanged.		
4.11 and Schedule 2	4.6	Slight amendment: the clause repeats the text of MBW 1998 with slight changes that improve the drafting or syntax, but the effect is unchanged.		
4.11 and Schedule 2 (paragraph 13)	4.6	Slight amendment: the clause repeats the text of MBW 1998 with slight changes that improve the drafting or syntax, but the effect is unchanged.		
5.3.2 Contractor's Insurance	6.2	Slight amendment: the clause repeats the text of MBW 1998 with slight changes that improve the drafting or syntax, but the effect is unchanged.		
5.4A and 5.4B		Slight amendment: the clause repeats the text of MBW 1998 with slight changes that improve the drafting or syntax, but the effect is unchanged.	This section of the Contract Particulars seeks a statement as to which insurance provision is to be applied. In this instance the Parties are offered the option of striking out the provisions – if neither were to be struck out, then both insurances would be required.	
5.4A.1 and 5.4B.1	6.3A	**Major change:** the amendment changes a fundamental process or procedure of the contract compared to MBW 1998.	A significant improvement that provides that the Contractor must insure a default value of 15% of the value of an insured event to cover the cost of professional fees. If the Parties think a different percentage is more appropriate then they must enter this in the Contract Particulars.	Contractors should discuss the Employer's requirements in regard to professional fees with their insurers and Employers should consider whether 15% is a sufficient percentage to cover likely additional professional fees.

7.2 Adjudication	Article 6, 8.1, Supplemental Condition D2	**Major change:** the amendment changes a fundamental process or procedure of the contract compared to MBW 1998.	The process for the appointment of an adjudicator has been changed. MW 2005 provides for the Parties to choose one of the five named Adjudicator Nominating Bodies (ANB), who will then nominate an adjudicator. If this section of the Contract Particulars is not completed then the default position is that the Party referring the dispute may require one of the listed ANBs to make a nomination. In the absence of a solution by the Parties to a MBW 98, the default position was that the President or Vice-President of the RIBA would be ANB. When assembling the Contract Particulars before execution of the contract, the Parties may agree to name their own adjudicator or specify a preferred ANB to put forward an adjudicator for any disputes referred to adjudication.	When the Contract Particulars are being assembled the Parties should decide whether an adjudicator for the whole project should be identified or whether one of the listed ANBs should be chosen in preference to the others names. The Parties may choose to add an ANB that is not listed.
Schedule 1 and Schedule 2 Base Date	Supplemental Condition A4.6.1	**Major change:** the amendment changes a fundamental process or procedure of the contract compared to MBW 1998.	MBW 1998 placed this date 10 days before the date of the Agreement, whereas MW 2005 requires the Parties to define the date themselves in the Contract Particulars.	State the Base Date for the contract.
Schedule 1 Arbitration	Article 7A	**Major change:** the amendment changes a fundamental process or procedure of the contract compared to MBW 1998.	The default position for MW 2005 is that arbitration does not apply. This means that the Parties will rely on the courts and adjudicators if it is not possible to resolve a dispute by other means. If arbitration is preferred and the Parties	The merits of mandatory arbitration should be considered, and if it is required the Contract Particulars must be amended.

MW 2005 clause, reference or title	MBW 1998 clause, reference or title	Summary of change	Consequence of change and comments	New actions
			wish it to be the default dispute resolution process for all disputes arising out of the contract then this must be stated at this point in the Contract Particulars. The only other way for arbitration to be introduced is for both Parties to agree to appoint an arbitrator for a particular dispute an ad hoc basis. Note that the Parties may have recourse to adjudication irrespective of their selection in this clause. This means that a dispute may be adjudicated whether or not the court or an arbitrator is seized of it. It is also open to the Parties to adopt an alternative method for resolving disputes such as mediation (as is recommended in clause 7.1 of MW 2005).	
Attestation	Attestation page headed 'as witness the hands of the Parties hereto'	**Significant change:** the clause repeats most of the text of MBW 1998, but this is amended in a way that does affect the operation of the clause.	The attestation process is unchanged, but the words used have been updated and the 'Note on Execution' in MW 2005 is clearer than in MBW 1998. There is now space for witnesses to state their name and address in addition to their signature – perhaps it is thought that this will result in some level of increased probity in the execution of the documents. Like MBW 1998, MW 2005 provides attestation forms for limited companies and individuals. The guidance note points out that the attestation may not be correct for housing associations, partnerships and foreign companies, but this is not stated in the terms of the contract itself.	Witnesses to provide name and address as well as a signature.

CONDITIONS				
Section 1: Definitions and Interpretation			MW 2005 provides a schedule of the definitions used in the contract, whereas MBW 1998 did not.	
Definitions				
Agreement		**New definition:** this clause provides a new definition of a term or of a process, but does not cause a change to the rights or obligations of the Parties.	The term 'Agreement' is defined by reference to the Articles of Agreement.	
Article		**New definition:** this clause provides a new definition of a term or of a process, but does not cause a change to the rights or obligations of the Parties.		
Business Day		**New definition:** this clause provides a new definition of a term or of a process, but does not cause a change to the rights or obligations of the Parties.	The definition 'Business Day' is new, however the reckoning of periods of days (clause 1.4) and definition of 'Public Holiday' are unchanged. This is relevant because many of the time-periods stated in the contract make reference to 'days', not 'Business Days'. In consequence users should note which units of time they are dealing with as there are inevitably consequences for failing to get it right.	Note which kind of 'days' you are dealing with.
Conditions		**New definition:** this clause provides a new definition of a term or of a process, but does not cause a change to the rights or obligations of the Parties.		
Construction Industry Scheme (CIS)	5.3	**New definition:** this clause provides a new definition of a term or of a process, but does not cause a change to the rights or obligations of the Parties.		

285

MW 2005 clause, reference or title	MBW 1998 clause, reference or title	Summary of change	Consequence of change and comments	New actions
Contract Particulars	Information in various locations	**New definition:** this clause provides a new definition of a term or of a process, but does not cause a change to the rights or obligations of the Parties.	This is a new definition describing the Contract Particulars and also noting that they are completed by the Parties.	
Excepted Risks		**Major change:** the amendment changes a fundamental process or procedure of the contract compared to MBW 1998.	The creation of 'Excepted Risks' is new for MW 2005 and was not found in MBW 1998. The Excepted Risks are (in broad terms) 'nuclear episodes' or damage caused by speeding aircraft. Such incidents are excluded from the definition of 'Specified Perils'. This means that losses from these causes do not have to be insured under the terms of this contract. Whilst it would also appear that the occurrence of one of the Excepted Risks might not give rise to a right for either Party to terminate the contract (under clause 6.10.3) it may still be possible to demonstrate that the particular episode is such as to be force majeure (clause 6.10.1.1). Also, given their nature, such Excepted Risks may well be covered by 6.10.1.4 (Terrorism) or 6.10.1.5 (Government action).	The Parties should obtain insurance from brokers or insurance companies with experience of insuring construction works, and particularly works executed under JCT contracts. Discussions should be held with providers of insurance so that the policies purchased are compatible with the contract.
Fluctuations Option	4.6	**New definition:** this clause provides a new definition of a term or of a process, but does not cause a change to the rights or obligations of the Parties.	The fluctuations percentage itself must be stated in the Contract Particulars. There is a place to insert the figure – if there is no figure, there is no percentage uplift.	At the time of writing, inflation is stable at a relatively low level, and consequently such fluctuation clauses are not always needed. If the Parties consider that inflation is likely to increase, then the fluctuation clauses may become more significant.

				Either way, if an uplift is considered appropriate a percentage figure should be stated.
Health and Safety Plan	5.7	**New definition:** this clause provides a new definition of a term or of a process, but does not cause a change to the rights or obligations of the Parties.	The plan referred to in this definition is the same as the one referred to in MBW 1998.	
Insolvent	7.2.2	**New definition:** this clause provides a new definition of a term or of a process, but does not cause a change to the rights or obligations of the Parties.	The definition is itself found in clause 6.1 and is different to that found in MBW 1998 – see the notes under clause 6.1 for an explanation of the difference and the consequences.	
Interest Rate	4.2.2, 4.5.2	**New definition:** this clause provides a new definition of a term or of a process, but does not cause a change to the rights or obligations of the Parties.	A single rate of interest has been defined for the whole contract, rather than each clause requiring the statement of an Interest Rate as was the case under MBW 1998. Note that the Interest Rate of 5% above the Bank of England Base Rate remains the same as that under MBW 1998.	
Parties		**New definition:** this clause provides a new definition of a term or of a process, but does not cause a change to the rights or obligations of the Parties.		
Party		**New definition:** this clause provides a new definition of a term or of a process, but does not cause a change to the rights or obligations of the Parties.		
Provisional Sum	3.8	**New definition:** this clause provides a new definition of a term or of a process, but does not cause a change to the rights or obligations of the Parties.	This is a fairly simple definition of what a Provisional Sum is. The statement of a definition adds some value and underlines the point that there is not an obligation on the Employer compelling it to instruct Provisional Sums.	

287

MW 2005 clause, reference or title	MBW 1998 clause, reference or title	Summary of change	Consequence of change and comments	New actions
Public Holiday	1.6.2	No change: the clause repeats the text of MBW 1998.		
Recitals		**New definition:** this clause provides a new definition of a term or of a process, but does not cause a change to the rights or obligations of the Parties.		
Rectification Period	Defects Liability Period	**New definition:** this clause provides a new definition of a term or of a process, but does not cause a change to the rights or obligations of the Parties.	A change of the name of the period for correcting defects. The name has changed from 'Defects Liability Period' to 'Rectification Period'.	Update any standard documentation to reflect new terminology when using MW 2005.
Scheme		**Major change:** the rights of the Parties have been changed by this amendment compared to MBW 1998.	The reference to the Scheme is new, as MBW 1998 made use of JCT's own adjudication rules.	
Site Materials		**New definition:** this clause provides a new definition of a term or of a process, but does not cause a change to the rights or obligations of the Parties.		
Specified Perils	6.3A, 6.3.B	**New definition:** this clause provides a new definition of a term or of a process.	The scope of the clause has been slightly widened, e.g. the mechanism for the escape of water is no longer prescribed in the definition of 'Specified Perils'. In addition, the language has been updated – e.g. 'tempest' is no longer a Specified Peril, presumably on the grounds that 'storm' is now thought to adequately describe the same Specified Peril.	
Statutory Requirements	5.1	**Significant change:** the clause repeats most of the text of MBW 1998, but this is amended in a way that does affect the operation of the clause.	This is an updated version of the reference found in MBW 1998 clause 5.1. The definition does appear to have been widened, or at least has been made more explicit, by the addition of the reference to	

			regulations and bye-laws of local authorities, statutory undertakers etc.	
VAT	5.2, Supplemental Condition B1	**New definition:** this clause provides a new definition of a term or of a process, but does not cause a change to the rights or obligations of the Parties.		
Interpretation				
1.2 Agreement etc. to be read as a whole	3.6	**Significant change:** the clause repeats most of the text of MBW 1998, but this is amended in a way that does affect the operation of the clause.	The provisions of the second part of clause 3.6 of MBW 1998 have been imported (without any change of consequence). Note the addition of a term requiring the Agreement to be read as a whole – this is of value to the interpretation and construction of the terms of the contract.	
1.3 Headings, reference to persons, legislation etc.		**New definition:** this clause provides a new definition of a term or of a process but does not cause a change to the rights or obligations of the Parties.	This new clause might be termed a collection of legal 'boiler plate' that ensures that words are not taken out of context. The clause does not create any new obligations but is of value in supporting the operation of the contract.	
1.4 Reckoning of periods of days	1.6	Slight amendment: The clause repeats the text of MBW 1998 with slight changes that improve the drafting or syntax, but the effect is unchanged.		
1.5 Contracts (Rights of Third Parties) Act 1999	1.8	Slight amendment: The clause repeats the text of MBW 1998 with slight changes that improve the drafting or syntax, but the effect is unchanged.		
1.6 Giving or service of notices and other documents	1.5	Slight amendment: The clause repeats the text of MBW 1998 with slight changes that improve the drafting or syntax, but the effect is unchanged.		
1.7 Applicable law	1.7	Redraft	The new clause has the same result but is more concise.	

MW 2005 clause, reference or title	MBW 1998 clause, reference or title	Summary of change	Consequence of change and comments	New actions
Section 2: Carrying out the Works				
2.1 Contractor's Obligations	1.1, 5.1	**Significant amendment:** the clause has been significantly reworded and gives additional clarity, but does not give rise to new obligations.	The clause has been reworded and restructured. It has gained from being more directly stated and divided into sub-clauses. The obligation for the Contractor to act with 'due-diligence' is no longer stated. Note, however, that MW 2005 does make provision for failure to complete on time and states that failure to proceed regularly and diligently with the Works is a ground for the Employer to terminate the contract (clause 6.4.1.2). An express obligation to comply with and give notices required by the Statutory Requirements is stated here – this had previously been found in clause 5.1 of MBW 1998.	
2.2 Commencement and Completion	2.1	Slight amendment: the clause repeats the text of MBW 1998 but is amended to incorporate or accommodate changes introduced in other clauses.	The Parties state the dates of commencement and completion in the Contract Particulars – MBW 1998 provided a space in this clause for this information.	Complete the Contract Particulars.
2.3 Architect/Contract Administrator's duties	1.2	No change: the clause repeats the text of MBW 1998.		
2.4 Correction of inconsistencies	3.6	No change: the clause repeats the text of MBW 1998.	Note that the provision for the correction of inconsistencies and those stating the primacy of the Conditions were previously found together, but MW 2005 has relocated the latter to clause 1.2.	

			Note the new definition of Work Schedules is used.	
2.5 Divergences from Statutory Requirements	5.1		Slight amendment: the clause repeats the text of MBW 1998 with slight changes that improve the drafting or syntax, but the effect is unchanged.		
2.6 Fees or charges legally demandable			Slight amendment: the clause repeats the text of MBW 1998 with slight changes that improve the drafting or syntax, but the effect is unchanged.	Note that fees and charges are not reimbursable by the Employer unless this is expressly agreed.	Contractors should ensure that they have included for the payment of fees and charges within the agreed Contract Sum.
2.7 Extension of time	2.2		No change: the clause repeats the text of MBW 1998.	Save for the use of some of the new defined terms, the clause is identical to clause 2.2 of MBW 1998.	
2.8 Damages for non-completion	2.3		Slight amendment: the clause repeats the text of MBW 1998 with slight changes that improve the drafting or syntax, but the effect is unchanged.	If it is the intention of the Parties to include damages for non-completion as a part of the bargain, then both Parties should ensure that the Contract Particulars are completed correctly.	
2.9 Practical completion	2.4		No change: the clause repeats the text of MBW 1998.		
2.10 Defects	2.5		**Significant amendment:** the clause repeats most of the text of the MBW 1998, but this is amended in a way that does affect the operation of the clause.	The clause is unchanged save in one significant respect. Damage caused by frost before practical completion is no longer specifically referenced. Clause 5.1 of MBW 1998 required that damage caused by such frost was to be made good by the Contractor at the Contractor's expense. The change to MW 2005 suggests that the Contractor would not automatically be liable for loss caused by frost occurring before practical completion.	

MW 2005 clause, reference or title	MBW 1998 clause, reference or title	Summary of change	Consequence of change and comments	New actions
			Note that the Contractor is liable under this clause for damage if this was due to deficiencies in materials or to workmanship not in accordance with the contract and in such circumstances frost could be the mechanism that exposes the deficiency.	
2.11 Certificate of making good	2.5	No change: the clause repeats the text of MBW 1998.		If it is the intention of the Parties to include a period that exceeds three months for the rectification of defects as a part of the bargain then both Parties should ensure that the Contract Particulars reflect this.
Section 3: Control of the Works				
3.1 Assignment	3.1	Slight amendment: the clause repeats the text of MBW 1998 with slight changes that improve the drafting or syntax, but the effect is unchanged.	The improvement in the drafting has the effect of making the clause a more complete prohibition against assignment.	
3.2 Person-in-charge	3.3	No change: the clause repeats the text of MBW 1998.		
3.3 Sub-letting	3.2	**Major change:** the amendment changes a fundamental process or procedure of the contract compared to MBW 1998.	Sub-clause 3.3 is a new addition to the Minor Works form that provides for the termination of any sub-contracts that the Contractor may have entered into on the termination of the main contract.	To give proper effect to sub-clause 3.3, the Contractor should ensure that the sub-contract includes a provision that terminates the sub-contract as soon as the main contract is terminated.

3.4 Architect/Contract Administrator's Instructions	3.5	**Significant change:** the clause repeats most of the text of MBW 1998, but this is amended in a way that affects the operation of the clause.	The Contractor is now obliged to 'forthwith comply with' an instruction, whereas MBW 1998 required that instructions were something that the Contractor had to 'forthwith carry out'. Although this would seem to be a drafting change, there would appear to be a distinction in effect: Instructions under MW 2005 appear to have a wider scope than under MBW 1998 as a result of the change. Accordingly, MW 2005 caters for a class of instruction that does not require a positive action or that may demand compliance or observance even if there is no action to carry out or action to be deferred.	
3.5 Non-compliance with instructions	3.5	**Significant change:** the clause repeats most of the text of MBW 1998, but this is amended in a way that does affect the operation of the clause.	The change relates to the situation where the Contractor has failed to comply with the Architect/ Contract Administrator's instruction. Under MW 2005 the Employer may then instruct another Contractor to carry out 'any work whatsoever which may be necessary to give effect to the instruction', whereas MBW 1998 is more limited and refers only to another Party carrying out 'the work' that was subject to the unimplemented instruction. There is also a change in emphasis in the recoverability of the cost. The MBW 1998 provided that 'all cost incurred thereby' might be recovered. The MW 2005 provides that the recovery is only in respect of additional cost in connection with giving effect to the instruction. In this instance, MW 2005 is the narrower clause. The redrafted clause is therefore at once broader in the scope of work that can be given to another contractor following the	

293

MW 2005 clause, reference or title	MBW 1998 clause, reference or title	Summary of change	Consequence of change and comments	New actions
			Contractor's failure to comply with an instruction, but it is also more precise in detailing what cost can be recovered from the Contractor. Consequently, the Contractor's refusal does not place a restriction on the Employer's ability to have the work done; but neither is the refusal a blank cheque for the Employer to do the works at the Contractor's utter expense.	
3.6 Variations	3.7	Clause redrafted: the clause has been significantly reworded, but does not give rise to new obligations.	The clause has been redrafted into three sub-clauses. Clause 3.6.1 is unchanged and 3.6.3 is very similar to its predecessor, clause 3.7 of MBW 1998. The change is found in 3.6.2 where the contract now provides that the Contractor and Architect/Contract Administrator are obliged to endeavour to agree a price for work carried out pursuant to an instructed variation. MBW 1998 accommodated an agreement of the sums due for varied works but did not place an express obligation on the Parties to seek to do so, nor was there any suggestion (as there is in MW 2005) that the agreement should be reached before the instruction was carried out. The contract provides for the Architect/Contract Administrator to price the variation if agreement cannot be reached. Unless a Party actively sought to avoid reaching any kind of agreement relating to an instructed agreement, then this new provision should be of limited effect.	The Architect/Contract Administrator and the Contractor might find value in setting up and agreeing a variation procedure at the beginning of the project. The variation procedure would address issues like the pre-agreement of variations and time limits for the provision of quotations from the Contractor and acceptance by the Employer or the Architect/Contract Administrator.

3.7 Provisional Sums	3.8	Slight amendment: the clause repeats the text of MBW 1998 with slight changes that improve the drafting or syntax, but the effect is unchanged.		The variation procedure proposed above should be designed to cope with the instruction and valuation of works associated with Provisional Sums.
3.8 Exclusion from the Works	3.4	Slight amendment: The clause repeats the text of MBW 1998 with slight changes that improve the drafting or syntax, but the effect is unchanged.	There is a very slight change in that MW 2005 refers to a power to exclude from 'the site' as opposed to 'the Works', but the clause is otherwise unchanged. The opportunities for this change to have practical impact are considered to be limited.	
3.9 CDM Regulations – Undertakings to comply	5.6, 5.7, 5.9	Clause redrafted: the clause has been significantly reworded but does not give rise to new obligations.		
3.10 Appointment of successors	1.3, 5.8	Slight amendment: the clause repeats the text of MBW 1998 with slight changes that improve the drafting or syntax, but the effect is unchanged.		
Section 4: Payment				
4.1 VAT	5.2	Clause redrafted: the clause has been significantly reworded, but does not give rise to new obligations.		
4.2 Construction Industry Scheme (CIS)	4.1, 5.3	**Major change:** the amendment changes a fundamental process or procedure of the contract compared to MBW 1998.	The contract now makes direct reference to the application of the provisions of the Construction Industry Scheme. MBW 1998 applied the provisions of a supplemental provision.	The Employer should consult with his accountants or other professional advisors to be certain as to whether he will be considered to be a 'Contractor'. Should the advice confirm 'Contractor' status, then he should ensure that the Contractor confirms that the CIS requirements will be applied.

MW 2005 clause, reference or title	MBW 1998 clause, reference or title	Summary of change	Consequence of change and comments	New actions
4.3 Progress payments and retention	4.2	No change: the clause repeats the text of MBW 1998.	The clause is now subdivided, but is otherwise identical.	
4.4 Failure to pay amount due	4.2.2	Slight amendment: the clause repeats the text of MBW 1998 with slight changes that improve the drafting or syntax, but the effect is unchanged.	The clause provides that interest is to be paid in the event of late payment of sums due under clause 4.3. If the Contractor were due sums under any clause other than 4.3, this clause would not apply to provide interest in respect of late payment of those sums. MBW 1998 clause 4.2.2 was of general application and did not apply such a restriction; indeed, it was stated to apply to any sum due to the Contractor under the Agreement.	
4.5 Penultimate certificate	4.3	Slight amendment: the clause repeats the text of MBW 1998 with slight changes that improve the drafting or syntax, but the effect is unchanged.		
4.6 Notices of amounts to be paid and deductions	4.4.1 - 3	**Significant change:** the clause repeats most of the text of MBW 1998, but this is amended in a way that does affect the operation of the clause.	The scope of the certificates and notices required under this clause are better stated under MW 2005 than they were under MBW 1998. The first certificate that the Employer sends to the Contractor (under clause 4.6.1) may, in addition to being effective as a payment notice, also be effective as a withholding notice on its own. If this is the purpose of the notice, then the notice issued under 4.6.1 must include sufficient detail in terms of the sums being withheld and the reasons for the withholding. Clause 4.6.2 states no precondition in terms of the Employer's ability to issue a	Employers must ensure that they formulate their notices correctly and provide the relevant information in all instances, but especially where they propose combining the notices.

			withholding notice, thus an effective withholding notice can be given under clause 4.6.2 even if there is no payment notice under clause 4.6.1.	
4.7 Contractor's right of suspension	4.8	Slight amendment: the clause repeats the text of MBW 1998 with slight changes that improve the drafting or syntax, but the effect is unchanged.		
4.8 Final certificate				
4.8.1	4.5.1.1	Slight amendment: the clause repeats the text of MBW 1998 with slight changes that improve the drafting or syntax, but the effect is unchanged.		
4.8.2	4.5.1.2	**Significant change:** the clause repeats most of the text of MBW 1998, but this is amended in a way that does affect the operation of the clause.	The Employer is required to provide more information in the notice concerning the amount to be paid in the final certificate than it was required to under MBW 1998. MBW 1998 required the Employer to state what he would pay, but under MW 2005 there is the additional requirement to state what the amount he proposes to pay relates to and how it has been calculated. As with the payment procedure for interim payments (clause 4.6.2), if the certificate required by this clause were to be provided with sufficient information, then it could be a valid withholding notice. To be effective as a withholding notice, the certificate would have to state the sums being withheld and the reasons for the withholding.	Additional information to be provided with the final certificate. Again, as with notices associated with interim payments, Employers must ensure that they formulate their notices correctly and provide the relevant information in all instances, but especially where they propose combining the notices.
4.8.3	4.5.1.3	Clause redrafted: the clause has been significantly reworded, but does not give rise to new obligations.	There is no requirement for a certificate to be issued under clause 4.8.2 before a certificate can be issued under 4.8.3.	

297

JCT 2005 Minor Works Building Contract (MW 2005)

MW 2005 clause, reference or title	MBW 1998 clause, reference or title	Summary of change	Consequence of change and comments	New actions
4.8.4	4.5.1.4	Clause redrafted: the clause has been significantly reworded, but does not give rise to new obligations.	The Employer must pay the sum stated in the Final Certificate unless he has given a notice under either 4.8.2 or under 4.8.3. The Employer might give notice under both clauses.	
4.9 Failure to pay final amount	4.2.2, 4.5.2	Slight amendment: the clause repeats the text of MBW 1998, but is amended to incorporate or accommodate changes introduced in other clauses.	The right to interest on late payment against the final certificate is now stated here in a single clause, rather than being set out separately.	If a contractor seeks to enforce its right to interest on late payment then it is advisable to state the particular clause that gives rise to the right in the given instance.
4.10 Fixed price	4.7	Slight amendment: the clause repeats the text of MBW 1998 with slight changes that improve the drafting or syntax, but the effect is unchanged.		
4.11 Contribution, levy and tax changes	4.6	Slight amendment: the clause repeats the text of MBW 1998, but is amended to incorporate or accommodate changes introduced in other clauses.		
Section 5: Injury, Damage and Insurance				
5.1 Liability of the Contractor – personal injury or death	6.1	No change: the clause repeats the text of MBW 1998.	Note that the scope of the insurance required is now stated in clause 5.3.	
5.2 Liability of the Contractor – injury or damage to property	6.2	Slight amendment: the clause repeats the text of MBW 1998 with slight changes that improve the drafting or syntax, but the effect is unchanged.	Note that the scope of the insurance required is now stated in clause 5.3.	

5.3 Contractor's insurance of his liability	6.1, 6.2	Clause redrafted: the clause has been significantly reworded but does not give rise to new obligations.	The clause provides that the Contractor must insure his indemnities to the Employer – after all, what use is an indemnity if the indemnifying Party has no assets? Although redrafted, the provisions required of any insurance policy are almost unchanged. However, there is a slight change in the scope of insurance required in respect of personal injury or death. MBW 1998 required insurance in respect of 'any person under a contract of service or apprenticeship', whereas MW 2005 requires this insurance in respect of 'any employee' and does not define this term. In response to this amendment, Parties may wish to consult their insurers to determine whether their apprentices are employees and establish the status of self-employed staff.	Check insurance status of apprentices and self-employed staff.
5.4A Insurance of the Works by the Contractor	6.3A	Clause redrafted: the clause has been significantly reworded, but does not give rise to new obligations.	Whether or not this clause applies will depend upon whether the Parties select it in the Contract Particulars – it would tend to be required for new build construction works rather than refurbishment works. In addition, the Contract Particulars should state the percentage that should apply to cover professional fees in respect of correcting damage covered by the policy. The list of events has been replaced by the defined term 'Specified Perils'; the list found under that definition has itself been modernised (see Definitions).	Ensure that the correct option is selected in the Contract Particulars.
5.4B Insurance of the Works and any existing structures by the Employer	6.3B	Clause redrafted: the clause has been significantly reworded but does not give rise to new obligations.	Whether or not this clause applies will depend upon whether the Parties select it in the Contract Particulars – it would tend to be required for refurbishment works rather than new build construction works.	Ensure that the correct option is selected in the Contract Particulars.

MW 2005 clause, reference or title	MBW 1998 clause, reference or title	Summary of change	Consequence of change and comments	New actions
			In addition, the Contract Particulars should state the percentage that should apply to cover professional fees in respect of correcting damage covered by the policy. The list of fateful events has been replaced by the defined term 'Specified Perils', which has itself been modernised.	
5.5 Evidence of Insurance	6.4	**Significant change:** the clause repeats most of the text of MBW 1998, but this is amended in a way that does affect the operation of the clause.	The departure from MBW 1998 only concerns instances where the Employer is insuring the Works (under clause 5.4B). Firstly, the obligation to give evidence of insurance will not apply to Employers who are local authorities. Secondly, if evidence of insurance is required, the Contractor may insist upon 'documentary evidence;' this is a more prescriptive statement of what is to be provided – MBW 1998 required 'such evidence as the Contractor may reasonably require'.	
Section 6: Termination			The termination clause has been thoroughly reworked and new processes introduced. Also, there has been a change in terminology – the Contract deals with 'termination' rather than 'determination'.	
6.1 Meaning of insolvency	7.2.2, 7.3.2	**Major change:** the amendment changes a fundamental process or procedure of the contract compared to MBW 1998.	The definition of 'insolvency' is stated here and has been reworded and has gained some clarity as a result. The new sub-clause 6.1.5 also improves its application to non-incorporated Parties, such as partnerships.	
6.2 Notices under section 6	7.1, 7.2.1, 7.2.2, 7.3.1, 7.3.2.	Slight change: the clause repeats the text of MBW 1998 but makes use of new defined terms and makes slight changes, but the effect is unchanged.	This clause has swept up some of the procedural elements and concentrated them in one clause, whereas previously these had been found and repeated in a number of clauses.	

6.3 Other rights, reinstatement				
6.3.1	7.2.4, 7.3.4	Slight amendment: the clause repeats the text of MBW 1998 with slight changes that improve the drafting or syntax, but the effect is unchanged.		
6.3.2		**Major change:** this is a wholly new clause for MW 2005.	This clause is a statement that the Parties may agree to a reinstatement. Though there may be practical and commercial value in a proposal for reinstatement, there is no legal compulsion provided by this clause of MW 2005. This clause is perhaps best described as a suggestion worth prudent consideration – it does no more than state the common law position. The Parties are not obliged to consent, or indeed respond, to a suggestion of reinstatement, either in this contract or in law. In essence, termination will be effective unless both the Parties agree, and certainly the terminating Party changing its mind after issuing its notice is insufficient on its own.	
6.4 Default by Contractor	7.2.1	Slight amendment: the clause repeats the text of MBW 1998 but makes use of new defined terms and makes slight changes, but the effect is unchanged.		
6.5 Insolvency of Contractor	7.2.2	Clause redrafted: the clause has been significantly reworded, but does not give rise to new obligations.		
6.6 Corruption	5.5	Slight amendment: the clause repeats the text of MBW 1998, but is amended to incorporate or accommodate changes introduced in other clauses.	In part this term reflects the usage of this form of contract by local authorities. In essence, if the Contractor commits a corruption offence, then the Employer may terminate any contract immediately.	

MW 2005 clause, reference or title	MBW 1998 clause, reference or title	Summary of change	Consequence of change and comments	New actions
6.7 Consequences of termination under clauses 6.4 to 6.6			This clause is new and provides for a series of actions by the Employer and his team following the Employer's termination of the Contractor's employment.	
6.7.1		**Major change:** the amendment changes a fundamental process or procedure of the contract compared to MBW 1998.	This new clause makes the Employer's right to possess the site after termination express, and as such should not be considered a new right so much as a clarification of rights previously provided by MBW 1998. It also provides for a right for the Employer to use all Site Materials, although the Employer is still obliged to seek third party consent where relevant (for example, where he proposes to use rented plant or site accommodation).	
6.7.2	7.2.3	Slight amendment: the clause repeats the text of MBW 1998 with slight changes that improve the drafting or syntax, but the effect is unchanged.		
6.7.3		**Major change:** the amendment changes a fundamental process or procedure of the contract compared to MBW 1998.	There is now an obligation on the Employer (that may be satisfied by the Architect/Contract Administrator) to compile an account that records the costs associated with the project at the end of a contract terminated due to the Contractor's default.	Employer or Contract Administrator to prepare an account following termination by the Employer.
6.7.4		**Major change:** the amendment changes a fundamental process or procedure of the contract compared to MBW 1998.	This clause states how the account required under 6.7.3 is to be compiled. Clause 6.7.3 states the rules to be applied to determine whether the Contractor owes money to the Employer, or vice versa.	

6.8 Default by the Employer	7.3.1	**Major change:** the amendment changes a fundamental process or procedure of the contract compared to MBW 1998.	Although largely similar, there is one change of significance, as one of the Contractor's grounds for termination has been removed. Specifically, the latter part of clause 7.3.2 of MBW 1998, which provided a right of termination in the event of unavailability of the premises. Perhaps this is because the events giving rise to such termination are also covered (and arguably better dealt with) as a suspension under clause 6.8.2 of the new MW 2005 form.
6.9 Insolvency of Employer	7.3.2, 7.3.3	**Significant change:** this is a new clause that provides a definition of a term or clarification of a process.	Clause 6.9.2 is new but does not create a new obligation; instead it makes the express provision that the Contractor is not obliged to work for an Employer who is Insolvent.
6.10 Termination by either Party		**Major change:** the amendment changes a fundamental process or procedure of the contract compared to MBW 1998.	This clause brings in a right of termination for either Party. The circumstances for such termination might be generalised as 'events that are beyond the control of the parties' that cause a continuous suspension of the Works for more than one month. For example, if an exercise of statutory authority to remove the power of local authorities to enter the specific kind of contract were brought into effect and caused suspension for one month, either Party could terminate. Similarly, an instance of inundation or fire causing delay of more than a month would enable either Party to act under this clause. Clause 6.10.2 states that the Contractor may not take advantage of a Specified Peril caused by its own negligence and terminate the Contract.

303

MW 2005 clause, reference or title	MBW 1998 clause, reference or title	Summary of change	Consequence of change and comments	New actions
6.11 Consequences of termination under 6.8 to 6.10	7.3.3	**Major change:** the amendment changes a fundamental process or procedure of the contract compared to MBW 1998.	The consequence itself remains unchanged: the account is drawn up and if the Employer is the culpable cause of the termination, then the Contractor may be entitled to recover direct loss or damage that has been caused by the termination. Loss or damage under clause 6.11.1.3 following termination will not be recoverable by the Contractor if the Contractor is the negligent cause of the Specified Peril. The specific changes to the process of MBW 1998 is that clause 6.11.2.2 of the contract now provides that the Contractor can submit the cost of goods he is legally obliged to pay for, whereas MBW 1998 only permitted this for materials on site. This provision is supported by clause 6.11.4, which improves upon its predecessor and provides that upon payment, such goods become the property of the Employer (it also states that retention may not be deducted from the sums due to the Contactor).	
Section 7: Settlement of Disputes				
7.1 Mediation	Footnote (v)	**Major change:** the amendment changes a fundamental process or procedure of the contract compared to MBW 1998.	The clause suggests that the Parties mediate any disputes that arise. This 'recommendation' that the Parties consider mediating their disputes has no binding force and the Parties may not be compelled by this clause to mediate or even give serious consideration to mediating a dispute.	Notwithstanding the lack of an obligation it is wise to give serious consideration to mediating a dispute. If the project is in its early stages there may be sound reasons for avoiding adversarial processes.

			Mediation may, in the particular circumstances of a dispute, be a sensible and cost-effective way to resolve a dispute. Certainly the courts regularly endorse mediation and are diverting many cases in this direction and penalising in costs those who do not consider it.	
7.2 Adjudication	8.1	**Major change:** the rights of the Parties have been changed by this amendment compared to MBW 1998.	The JCT Adjudication rules have been replaced in favour of the Scheme found in Part 1 of the *Scheme for Construction Contracts (England and Wales) Regulations 1998* ('the Scheme'). MW 2005 departs from the Scheme only in as far as it provides for the Parties to identify their own adjudicator or express a preference for a specific Adjudicator Nominating Body in the Contract Particulars. The Scheme is described in detail in other publications.	
7.3 Arbitration	8.2	Slight amendment: the clause repeats the text of MBW 1998 with slight changes that improve the drafting or syntax, but the effect is unchanged.	The arbitration rules are set out in Schedule 1 of MW 2005 – these make extensive reference to the 2005 version of the Construction Industry Model Arbitration Rules (CIMAR).	
SCHEDULES				
Schedule 1 Arbitration	Supplemental Condition E: Arbitration			
1. Conduct of arbitration	E1, E7	Slight amendment: the clause repeats the text of MBW 1998 with slight changes that improve the drafting or syntax, but the effect is unchanged		

305

MW 2005 clause, reference or title	MBW 1998 clause, reference or title	Summary of change	Consequence of change and comments	New actions
2. Notice of reference to arbitration	E2.1, E2.2, E2.3	Clause redrafted: the clause has been significantly reworded, but does not give rise to new obligations.	The new clause does not incorporate a quotation of the CIMAR, but does state application of the relevant requirements of those Rules.	
3. Powers of the Arbitrator	E3	No change: the clause repeats the text of MBW 1998.		
4. Effect of award	E4	No change: the clause repeats the text of MBW 1998.		
5. Appeal – questions of law	E5	Slight amendment: the clause repeats the text of MBW 1998 with slight changes that improve the drafting or syntax, but the effect is unchanged.		
6. Arbitration Act 1996	E6	Slight amendment: the clause repeats the text of MBW 1998 but makes use of new defined terms and makes slight changes but the effect is unchanged.		
Schedule 2 Fluctuations Option – Contribution, levy and tax changes				
1. Deemed calculation of Contract Sum – labour	A1 (A1–A1.9 inclusive)	Slight amendment: the clause repeats the text of MBW 1998 but makes use of new defined terms and makes slight changes, but the effect is unchanged.	The provisions of MBW 1998 are repeated virtually word-for-word, save for the use of new defined terms that slightly improve the comprehensibility of the terms. In addition, the reference to the *Social Security Pensions Act* 1975 has been updated to the *Pension Schemes Act* 1993.	

2. Deemed calculation of Contract Sum – materials	A2.1 -2	Slight amendment: the clause repeats the text of MBW 1998 but makes use of new defined terms and makes slight changes, but the effect is unchanged.	There are some small changes reflecting the use of defined terms, but these do not affect the application of the contract.	
3. Sub-let work – incorporation of provisions to like effect	A3.1	**Major change:** the rights of the Parties have been changed by this amendment compared to MBW 1998.	Where this Schedule applies, and where the Contractor sub-contracts works, then the Contractor is now required to incorporate the provisions of paragraph 2 into its sub-contracts (this was not a requirement of MBW 1998). Otherwise the text is the same and paragraph 3.2 is identical to A3.2.	The Contractor is to incorporate additional terms into sub-contracts.
4. Written notice by Contractor	A4.1	No change: the clause repeats the text of MBW 1998.		
5. Agreement – Architect/ Contract Administrator and Contractor	A4.3	Slight amendment: the clause repeats the text of MBW 1998 but makes use of new defined terms and makes slight changes, but the effect is unchanged.		
6. Fluctuations added to or deducted from Contract Sum	A4.4	**Major change:** the rights of the Parties have been changed by this amendment compared to MBW 1998.	Where this Schedule applies, then sums that are allowable under paragraph 2 will be added to or deducted from the Contract Sum. Sums added under this paragraph were not permitted to be added to the Contract Sum under MBW 1998.	
7. Evidence and computations by Contractor	A4.4.1	**Major change:** the rights of the Parties have been changed by this amendment compared to MBW 1998.	MBW 1998 required the Contractor to provide evidence – MW 2005 now states an additional express requirement for computations too.	
8. Actual payment by Contractor	A4.4.2	Slight amendment: the clause repeats the text of MBW 1998 with slight changes that improve the drafting or syntax, but the effect is unchanged.		
9. No alteration to Contractor's profit	A4.4.3	Slight amendment: the clause repeats the text of MBW 1998 with slight changes that improve the drafting or syntax, but the effect is unchanged.		

MW 2005 clause, reference or title	MBW 1998 clause, reference or title	Summary of change	Consequence of change and comments	New actions
10. Position where Contractor in default over completion	A4.4.4.1 - 2	Slight amendment: the text of MBW 1998 is repeated, but new defined terms are used, slight changes are made and the clause is amended to accommodate changes introduced in other clauses.		
11. Work etc to which paragraphs 1 to 3 not applicable	A4.5	Slight amendment: the text of MBW 1998 is repeated, but new defined terms are used, slight changes are made and the clause is amended to accommodate changes introduced in other clauses.	MW 2005 expands the exclusions of MBW 1998 to apply to materials under paragraph 2.	
12. Definitions				
12.1 Base Date	A4.6.1	**Major change:** the rights of the Parties have been changed by this amendment compared to MBW 1998.	MBW 1998 placed this date 10 days before the date of the Agreement, whereas MW 2005 requires the Parties to define the date themselves in the Contract Particulars.	Note relevance of Base Date to fluctuations when stating the Base Date in the Contract Particulars.
12.2 Materials and goods	A4.6.2	No change: the clause repeats the text of MBW 1998.		
12.3 'Workpeople'	A4.6.3	No change: the clause repeats the text of MBW 1998.		
12.4 'Wage-fixing body'	A4.6.4	Slight amendment: the clause repeats the text of MBW 1998 with slight changes that improve the drafting or syntax, but the effect is unchanged.	The reference to legislation has been removed and the reference to 'recognised terms and conditions' replaces it.	
12.5 Recognised terms and conditions		**Major change:** this is a wholly new clause for MW 2005.	This new definition seeks to describe a set of terms and conditions that have achieved a level of authority or recognition in the marketplace for the kind of works under consideration.	

			This is a difficult definition and it will be difficult to apply when ever used. It depends upon a number of subjective points about which Parties might reasonably assume different positions.	
13. Percentage addition to fluctuation payments or allowances	A5	**Major change:** the rights of the Parties have been changed by this amendment compared to MBW 1998.	The clause is the same as that in MBW 1998 save that the reference to clause A3.2 has been deleted. This means that the Contractor does not get a percentage added for sums it is obliged to pay under a sub-contract.	

Contract clause headings and numbering structure from the *Minor Works Building Contract*, Joint Contracts Tribunal Limited, 2005, Sweet and Maxwell, © The Joint Contracts Tribunal Limited 2005, are reproduced here with permission.

Table of destinations: MBW 1998 and MW 2005

This table of destinations follows the structure of MBW 1998 and states the new clause reference in the MW 2005 form.

MW 1998	MW 2005	MW 1998	MW 2005
Parties to Agreement	Parties to Agreement	1.6.2	1.1 (definition)
1st recital	First recital and Second recital	1.7	1.7
2nd recital	Third recital	1.8	1.5
3rd recital	Second recital	2.1	2.2
4th recital		2.2	2.7
5th recital	Fourth recital	2.3	2.8
Article 1	Article 1	2.4	2.9
Article 2	Article 2	2.5	2.10 and 2.11
Article 3	Article 3	3.1	3.1
Article 4	Article 4	3.2	3.3
Article 5	Article 5	3.3	3.2
Article 6	Article 6	3.4	3.8
Article 7A	Article 7 and Schedule 1	3.5	3.4 and 3.5
Article 7B	Article 8	3.6	1.2 and 2.4
Article 7C		3.7	3.6
Attestation	Attestation	3.8	3.7 and 1.1 (definition)
1.1	2.1	4.1	4.2
1.2	2.3	4.2.1	4.3
1.3	3.10	4.2.2	4.3, 4.4, 4.9 and 1.1 (definition)
1.4		4.3	4.5
1.5	1.6	4.4	4.6
1.6.1	1.4	4.5.1.1	4.8.1

MW 1998	MW 2005
4.5.1.2	4.8.2
4.5.1.3	4.8.3
4.5.1.4	4.8.4
4.5.2	1.1 (definition), 4.9
4.6	4.11, Schedule 2, 1.1 (definition)
4.7	4.10
4.8	4.7
5.1	2.1, 2.5 and 1.1 (definition)
5.2	4.1 and 1.1 (definition)
5.3	4.2 and 1.1 (definition)
5.4	
5.5	6.6
5.6	3.9
5.7	1.1 (definition), 3.9
5.8	3.10
5.9	3.9
6.1	5.1 and 5.3
6.2	5.2 , 5.3
6.3A	5.4A, 5.4B.1 and 1.1 (definition)
6.3B	5.4B and 1.1 (definition)
6.4	5.5
7.1	6.2
7.2.1	6.2 and 6.4
7.2.2	6.1, 6.2, 6.5 and 1.1 (definition)
7.2.3	6.7
7.2.4	6.3.1

MW 1998	MW 2005
7.3.1	6.2 and 6.8
7.3.2	6.1 and 6.2
7.3.3	6.9 and 6.9
7.3.4	6.3.1
8.1	7.2
8.2	7.3
8.3	
Supplemental Conditions	
A1	Schedule 2, para 1
A2	Schedule 2, para 2
A3	Schedule 2, para 3
A4.1	Schedule 2, para 4
A4.2	Schedule 2, para 4
A4.3	Schedule 2, para 5
A4.4	Schedule 2, para 6
A4.4.1	Schedule 2, para 7
A4.4.2	Schedule 2, para 8
A4.4.3	Schedule 2, para 9
A4.4.4	Schedule 2, para 10
A4.5	Schedule 2, para 11
A4.6	Schedule 2, para 12.1
A5	Schedule 2, para 13
B1	1.1 (definition)
B2 to B10	
C1 to C14	
D1	

MW 1998	MW 2005
D2	7.2
D3 to D8	
E1	Schedule 1, para 1
E2	Schedule 1, para 2
E3	Schedule 1, para 3

MW 1998	MW 2005
E4	Schedule 1, para 4
E5	Schedule 1, para 5
E6	Schedule 1, para 6
E7	Schedule 1, para 1

Index

323

327